Developing animal feed products

It is widely recognised that agriculture is a significant contributor to global warming and climate change. Agriculture needs to reduce its environmental impact and adapt to current climate change whilst still feeding a growing population, i.e. become more 'climate-smart'. Burleigh Dodds Science Publishing is playing its part in achieving this by bringing together key research on making the production of the world's most important crops and livestock products more sustainable.

Based on extensive research, our publications specifically target the challenge of climate-smart agriculture. In this way we are using 'smart publishing' to help achieve climate-smart agriculture.

Burleigh Dodds Science Publishing is an independent and innovative publisher delivering high quality customer-focused agricultural science content in both print and online formats for the academic and research communities. Our aim is to build a foundation of knowledge on which researchers can build to meet the challenge of climate-smart agriculture.

For more information about Burleigh Dodds Science Publishing simply call us on +44 (0) 1223 839365, email info@bdspublishing.com or alternatively please visit our website at www.bdspublishing.com.

Related titles:

Improving grassland and pasture management in temperate agriculture
Print (ISBN 978-1-78676-200-9); Online (ISBN 978-1-78676-202-3, 978-1-78676-203-0)

Improving gut health in poultry
Print (ISBN 978-1-78676-304-4); Online (ISBN 978-1-78676-306-8, 978-1-78676-307-5)

Improving rumen function
Print (ISBN 978-1-78676-332-7); Online (ISBN 978-1-78676-334-1, 978-1-78676-335-8)

Reducing greenhouse gas emissions from livestock production
Print (ISBN 978-1-78676-439-3); Online (ISBN 978-1-78676-441-6, 978-1-78676-442-3)

Chapters are available individually from our online bookshop:
https://shop.bdspublishing.com

climate•SMART•publishing
IN AGRICULTURAL SCIENCE

BURLEIGH DODDS SERIES IN AGRICULTURAL SCIENCE
NUMBER 101

Developing animal feed products

Edited by Dr Navaratnam Partheeban, formerly Royal Agricultural University, UK

burleigh dodds
SCIENCE PUBLISHING

Published by Burleigh Dodds Science Publishing Limited
82 High Street, Sawston, Cambridge CB22 3HJ, UK
www.bdspublishing.com

Burleigh Dodds Science Publishing, 1518 Walnut Street, Suite 900, Philadelphia, PA 19102-3406, USA

First published 2021 by Burleigh Dodds Science Publishing Limited

Library of Congress Control Number: 2021937774

British Library Cataloguing in Publication Data
A catalogue record for this book is available from the British Library

ISBN 978-1-78676-463-8 (Print)
ISBN 978-1-78676-466-9 (PDF)
ISBN 978-1-78676-465-2 (ePub)
ISSN 2059-6936 (print)
ISSN 2059-6944 (online)

DOI 10.19103/AS.2021.0083

Typeset by Deanta Global Publishing Services, Dublin, Ireland

Contents

Series list

Title	Series number
Achieving sustainable cultivation of maize - Vol 1 From improved varieties to local applications Edited by: Dr Dave Watson, CGIAR Maize Research Program Manager, CIMMYT, Mexico	001
Achieving sustainable cultivation of maize - Vol 2 Cultivation techniques, pest and disease control Edited by: Dr Dave Watson, CGIAR Maize Research Program Manager, CIMMYT, Mexico	002
Achieving sustainable cultivation of rice - Vol 1 Breeding for higher yield and quality Edited by: Prof. Takuji Sasaki, Tokyo University of Agriculture, Japan	003
Achieving sustainable cultivation of rice - Vol 2 Cultivation, pest and disease management Edited by: Prof. Takuji Sasaki, Tokyo University of Agriculture, Japan	004
Achieving sustainable cultivation of wheat - Vol 1 Breeding, quality traits, pests and diseases Edited by: Prof. Peter Langridge, The University of Adelaide, Australia	005
Achieving sustainable cultivation of wheat - Vol 2 Cultivation techniques Edited by: Prof. Peter Langridge, The University of Adelaide, Australia	006
Achieving sustainable cultivation of tomatoes Edited by: Dr Autar Mattoo, USDA-ARS, USA; and Prof. Avtar Handa, Purdue University, USA	007
Achieving sustainable production of milk - Vol 1 Milk composition, genetics and breeding Edited by: Dr Nico van Belzen, International Dairy Federation (IDF), Belgium	008
Achieving sustainable production of milk - Vol 2 Safety, quality and sustainability Edited by: Dr Nico van Belzen, International Dairy Federation (IDF), Belgium	009
Achieving sustainable production of milk - Vol 3 Dairy herd management and welfare Edited by: Prof. John Webster, University of Bristol, UK	010
Ensuring safety and quality in the production of beef - Vol 1 Safety Edited by: Prof. Gary Acuff, Texas A&M University, USA; and Prof. James Dickson, Iowa State University, USA	011
Ensuring safety and quality in the production of beef - Vol 2 Quality Edited by: Prof. Michael Dikeman, Kansas State University, USA	012
Achieving sustainable production of poultry meat - Vol 1 Safety, quality and sustainability Edited by: Prof. Steven C. Ricke, University of Arkansas, USA	013
Achieving sustainable production of poultry meat - Vol 2 Breeding and nutrition Edited by: Prof. Todd Applegate, University of Georgia, USA	014

Introduction

The animal feed sector faces increasingly complex challenges. It needs to improve feed digestibility/efficiency whilst also promoting growth and enhancing both product quality and safety. At the same time, it has an increasingly important role in helping to promote animal health, welfare and the sustainability of farming.

Part 1 of this volume reviews advances in optimising the key stages in developing successful new animal feed products, from assessing feed ingredients, product development and processing to maintaining quality and safety. Chapters also feature authoritative discussions by leading experts on the relationship between nutrition and animal health. Part 2 examines quality and safety assurance by demonstrating how best to implement risk management systems for the prevention and control of contaminants present in animal feed.

Part 1 Developing animal products

Part 1 opens with a chapter focused on techniques for identifying new animal feed ingredients and additives. The identification of potential novel feed ingredients requires quick and effective analytical techniques to assess their composition and functionality. Chapter 1 provides a practical guide to routine and standard methods to detect the main traits of interest in animal feed ingredients and additives. The chapter discusses techniques to evaluate feed nutritional value, including chemical composition and nutrient digestibility. It then reviews different *in vitro* methods for feed evaluation systems as well as assessment of nutrient molecular structure. The chapter also discusses analysis of feed bioactive compounds and their functionality. The chapter concludes with a case study on seaweed as a potential novel feed for livestock.

The next chapter reviews the effect of processing techniques on the quality of animal feed. Chapter 2 begins by examining how extrusion cooking can directly influence the nutritional value of a given formulation component, such as starch, in a dry animal feed formulation. It then goes on to discuss the processing of cereals for food and feed, focusing specifically on the range of cereal processing conditions currently used in the industry. A case study on developing animal feed products from starch is also provided, primarily focusing on dry cat food. The chapter then summarises the importance of extrusion cooking technology in the manufacturing of food and feed products and the need to achieve the correct level of water solubility in these products.

Expanding on topics previously discussed in Chapter 2, Chapter 3 focuses on the processing techniques that are used to optimise digestibility and the nutritional value of animal feed. It discusses the two most popular methods of particle size reduction, hammermills and roller mills. The chapter then moves on to review the effects of particle size reduction on swine and broiler feed production and performance, which is then followed by an analysis of the effects of pelleting on swine and broiler feed performance. The chapter concludes by highlighting the importance of feed processing research.

The final chapter of Part 1 addresses trends in analytical techniques for testing animal feed. The challenges for the feed analysis sector are to ensure feed quality and safety. Innovative strategies need to be implemented to authenticate feed and feed ingredients and to check that they fulfil all labelling obligations. Chapter 4 focuses on the authentication of feed by different but complementary analytical techniques, such as microscopy, vibrational spectroscopy, genomics and proteomics. The chapter explores ways techniques can be combined to find solutions for complex testing challenges, as well as developments in data science to achieve faster, real-time analysis.

Part 2 Quality and safety assurance

The first chapter of Part 2 focuses on the developments in techniques to test the efficacy of animal feed products. Animal feed products affect both healthy growth of livestock, their environmental impact (e.g. through emissions) and the safety and quality of livestock food products. All these areas are regulated by agencies such as the European Food Safety Authority (EFSA). Chapter 5 first discusses EFSA guidance on how to compile dossiers for feed additives. The chapter then discusses key issues and steps in demonstrating the efficacy of new animal feed products in key areas such as: reduction of nitrogen (N) excretion, reduction of feed contamination of by mycotoxins and, finally, reduction of methane emissions with feed additives. Key steps, such as the use of *in vitro* and *in vivo* studies, to test the effects of feed additives are discussed in detail.

Chapter 6 examines advances in understanding key contamination risks in animal feed, specifically mycotoxins. Mycotoxins accumulate in corn, cereals, soybeans, sorghum, peanuts, and other food and feed crops, directly in the field or during transportation, processing or storage. Consumption of mycotoxin-contaminated food or feed can lead acute or chronic toxicity in humans and animals, as well as crop losses. The chapter reviews the toxicity of the six mycotoxins, the foods they commonly contaminate, and the current methods used to detect and control these mycotoxins.

Chapter 7 considers risk management systems for prevention and control of contaminants in animal feed. Feed safety remains an issue for both the health of livestock and as a prerequisite for food safety. Risk assessment plays a role in identifying and determining potential risks of contaminants. When contaminants limits are defined, risk management systems can be designed for prevention and control. This chapter focuses on management systems used to prevent and control contaminants in animal feed. A brief introduction to feed contaminants and new feed ingredients is given, followed by a description of the main risk management systems applied in Europe and the US.

The final chapter of the book addresses developing effective product dossiers for regulatory approval of new animal feed products. Chapter 8 describes the key steps in compiling product dossiers required to gain regulatory approval for new animal feed products. As an example, the chapter describes the data needed for a feed additive dossier in the European Union (EU) as mandated by the EFSA.

Part 1

Developing animal products

Chapter 1

Techniques for identifying new animal feed ingredients and additives

Jordi Ortuño, Institute of Global Food Security - Queen's University Belfast, UK; Alexandros Ch. Stratakos, Centre for Research in Biosciences - University of the West of England, UK; and Katerina Theodoridou, Institute of Global Food Security - Queen's University Belfast, UK

1 Introduction

Food supply in many regions has never been so secure yet, at the same time, faced greater challenges. Global warming, the exponential rise of urban populations and globalisation (with the risk of crises such as COVID-19) are promoting competition between production of food, feed and fuel that will require innovative solutions to ensure enough safe food in the future. Animal food production is at the forefront of this challenge, reflected in the transition towards a more sustainable model of production. The global demand for improved food safety and animal welfare, together with reduced air-, water- and soil-derived contamination, is incompatible with many current production systems. As an example, the European Union, the major global producer of dairy products and second in production of pig meat, is only able to produce 20% of the protein requirement in livestock feed. This deficiency is made up

http://dx.doi.org/10.19103/AS.2020.0083.05

by imports of vast amounts of soy from other countries, often with a huge environmental impact derived from changes in land use, such as deforestation of vast regions of the Amazon rain forest in Brazil.

The search for new alternative feed resources must avoid direct or indirect competition with human food resources. Beyond that, ideal novel ingredients and additives should not only provide similar or better production performance than conventional ingredients, but also offer extra added value in addressing challenges such as improving animal health and welfare (e.g. in promoting gut health) or reducing contamination risks. A number of examples illustrate this.

One example is marine algae (micro and macro) which meet nutritional requirements for ruminants and, at the same time, do not compete for arable land and are less vulnerable to weather fluctuations related to climate change. They also possess bioactive compounds (bromoform, phlorotannins, saponins) capable of reducing ruminal methane production by up to 98%. A second example is insects, which can upcycle agri-food wastes into feed ingredients, potentially reducing reliance on cultivated soy. In addition, the presence of chitin in insect exoskeletons has prebiotic effects that promote beneficial bacteria in the gut, competitively excluding pathogens and promoting healthy gut function. Another example is silvopastoralism, a multifunctional system that combines animal production with pasture, shrubs and trees, thus promoting environmental protection and soil conservation. Most of the tree species employed in such systems are abundant in condensed tannins, which are able to reduce ammonia emissions, a key factor in soil and water acidification, eutrophication and global warming.

The identification of these potential novel ingredient sources requires rapid and effective analytical techniques capable of assessing composition and functionality, and screening for the effects of intrinsic variability (e.g. resulting from species or cultivars) and extrinsic variability (e.g. seasons, weather conditions, substrate composition). The present chapter aims to provide a practical guide to routine and standard methods employed to detect the main traits of interest in different sources of animal feed ingredients and additives.

2 Chemical analysis to determine protein, fibre, fatty acid content and digestibility

Assessing the nutritional value and quality of a new animal feed ingredient or additive for livestock diets requires determining chemical composition, intake, palatability, acceptability and digestibility. This section looks at ways of analysing the chemical composition of an ingredient or additive.

2.1 Protein content and digestibility

Crude protein content can be estimated by multiplying the nitrogen content by 6.25 to provide a quantitative indicator of feed protein content. This conversion factor has been used since the 19th century and assumes the nitrogen content of proteins to be 16% (Mariotti et al., 2008). However, this approach can be inaccurate in the case of novel ingredients (e.g. seaweed or insects), often resulting in an overestimation of protein content. Specific nitrogen-to-protein conversion factors, calculated from amino acid composition and total nitrogen, provide more reliable protein quantification.

Different methods are commonly used to determine feed protein content. The first is the classical Kjeldahl method for the determination of feed nitrogen (AOAC, 1990, method 977.02). The Dumas method is also widely used to assess crude protein content, is fast and requires no corrosive chemicals in contrast to the Kjeldahl method. The Dumas method measures all forms of nitrogen however provides a slight overestimation of total protein nitrogen content compared to the Kjeldahl method (Mæhre et al., 2018).

Identification of the protein value of a new feed requires assessment of not only protein quantity but also its quality. Protein quality can be determined partly by the quantity of essential and non-essential amino acids (AAs), and partly by their digestibility which measures the amount that can be absorbed from the gut and utilized. The 'traditional' method for the analysis of amino acids is based on ion-exchange chromatography with spectrophotometric detection (IEC-Vis) in which AAs are separated in the chromatographic column and subjected to post-column derivation with ninhydrin. IEC-Vis is recommended as a technique for official feed control in numerous countries.

The amino acid tryptophan can be determined by the high-performance liquid chromatography (HPLC) method with fluorometric detection (Szkudzińska et al., 2017). However, the analysis of AAs is complicated by the lack of a simple and reliable protein hydrolysis method. A simple, accurate and reliable method for the analysis of AA composition is Reversed-Phase High-Performance Liquid Chromatography with Diode-Array Detection (RP-HPLC-DAD), making analysis simple and accessible to any routine analytical laboratory (Liyanaarachchi et al., 2020).

Rapid and reproducible methods have been developed for the estimation of *in vitro* feed digestibility in ruminants and monogastric animals. There are several *in vitro* methods for assaying protein digestibility in monogastric animals which work by simulating the digestive process of the studied animal. The pepsin digestibility assay is one of the most widely used assays to evaluate the quality of feed and protein ingredients in poultry (Bryan and Classen, 2020).

In vivo animal trials are among the most accurate ways to measure protein digestibility in nutritional evaluation of feeds but such trials are time consuming and expensive. *In vivo* estimation of protein quality of a feed ingredient is normally achieved by feeding the ingredient to the intended animal while assessing the extent to which nutrients are absorbed by the terminal intestine.

In vitro methods have been developed to estimate protein digestibility of ingredients in pigs, simulating protein digestion along the gastrointestinal tract (GIT) by incubating protein sources with pepsin and pancreatin sequentially, using incubation conditions matching those in the GIT. One problem is that protein sources with a similar *in vitro* protein digestibility, based on determination of nitrogen solubility, can differ in the extent of release of free amino acids or di- and tri-peptides. Chen et al. (2018b) have developed a method for the prediction of *in vivo* protein digestion kinetics.

2.2 Fibre content

Fibre is characterised as hydrolytically indigestible and partially fermentable and is a complex and highly variable component of plant-based feedstuffs (NRC, 2007). The most ways to estimate fibre content in feed ingredients include (Kerr and Shurson, 2013):

- crude fibre (CF),
- acid detergent fibre (ADF),
- neutral detergent fibre (NDF),
- soluble and insoluble fractions of total dietary fibre (TDF), and
- non-starch polysaccharide.

The method for the determination of crude fibre is designed to divide carbohydrates into digestible and indigestible fractions. It is an indicator of the energy value of the feed: when crude fibre values are high, feed energy is low because crude fibre is considered indigestible. Crude fibre accounts for most of the cellulose and a portion of the hemicellulose and lignin but no ash and, as a result, underestimates true fibre. CF is not a good indicator of digestibility in ruminant animals, and the use of this assay in feeds for ruminants has declined significantly (Fahey et al., 2018).

To measure detergent fibres (ADF and NDF) the Van Soest 1963 detergent fibre analysis system is used. Plant cell substances can be divided into less digestible cell walls (hemicelluloses, cellulose and lignin) and highly digestible cell contents (starch and sugars). The acid detergent fibre (ADF) represents the least digestible fibre portion in the animal feed. The highly digestible cell contents are separated from the cell walls by using two different detergent systems:

- NDF = Hemicelluloses + Cellulose + Lignin + Ash, and
- ADF = Cellulose + Lignin + Ash.

ADF is the residue remaining after boiling a test material in an acid detergent solution. The neutral detergent fibre (NDF) is the residue or insoluble fraction left after boiling a feed material in a neutral detergent solution and contains insoluble plant cell wall components.

Non-starch polysaccharides (NSP), together with lignin, have been defined as the total dietary fibre (TDF) in feedstuffs. In the last few decades, there has been a great interest in gut microbiota and their effect on the host's metabolic regulation. A beneficial shift in the microbial ecosystem of animals can be promoted by various dietary and non-dietary interventions. Diet is ranked as one of the most important and effective regulators of gut microbiota composition, specifically dietary fibre. Studies with pigs and poultry have shown that fermentation characteristics of fibres and their beneficial effects on gut health vary widely, based on type, form and physicochemical properties of the TDF. It is important, therefore, to have information on the different TDF types in a new feed additive or ingredient (Jha et al., 2019). TDF is primarily analysed by enzymatic-gravimetric methods which solubilise the different fibres fractions with enzymes and solvents and measure the weight of residues after these treatments (Gidenne, 2014).

2.3 Fatty acid content

Fatty acids are carboxylic acids with either saturated or unsaturated aliphatic chains. They can be divided according to chain length into (Baltić et al., 2017):

- short-chain fatty acids (<C6),
- medium-chain fatty acids (C6-C12),
- long-chain fatty acids (C13-C21), and
- very-long-chain fatty acids (≥C22)

Fatty acids can be present in their free fatty acid (FFA) forms. However, they most often exist in bound forms such as cholesterol and phospholipids (PL). In feeds, lipids are mainly found in the form of triacylglycerols (triacylglycerides - TAG), which make up to 99% of lipids of plant and animal origin (Fennema, 1996).

The diet of animals influences the fatty acid composition of animal products (i.e. meat, milk and eggs). This relationship is stronger in monogastrics than in ruminants where dietary fatty acids are hydrogenated in the rumen (Chilliard et al., 2008). There is an increasing recognition of the health benefits of polyunsaturated fatty acids (PUFA) which are divided into omega-3 and omega-6 fatty acids as well as docosahexaenoic acid (DHA). These are seen as having nutraceutical properties e.g. in the development of the nervous system and the maintenance of cognitive function (Butler, 2014; Balić et al.,

2020). The inclusion of fats in ruminant diets is also considered a promising way to manipulate the rumen microbial community to reduce methane emissions (Vargas et al., 2020). It is therefore important for human and animal health as well as the environment to determine the fatty acid content of a potential new animal feed ingredient or additive before use in animal nutrition.

Total lipid content of feedstuffs is generally estimated by the Soxhlet method or by cold extraction methods using organic solvents (Bligh and Dyer, 1959 and Folch et al., 1957 techniques). Both methods rely on chloroform and methanol to form a monophasic solvent system to extract and dissolve the lipids. The Folch method is regarded as the most reliable for complete recovery of total lipids, but the Bligh and Dyer method is more widely known and used. After transmethylation the fatty acid composition of the extracted oil is determined by analytical techniques such as gas-liquid chromatography, mass spectrometry (GC-MS), gas chromatography with flame ionization detection (GC-FID), and liquid chromatography-mass spectrometry (LC-MS) (Leiva and Granados-Chinchilla, 2020; Chiu and Kuo, 2020).

3 Techniques to determine nutritive value and digestibility: rumen *in vitro* fermentation techniques, protein evaluation systems and models

As discussed in the last section, the nutrient composition of feed is commonly determined in the lab with chemical analysis. However, this is not sufficient to determine its nutritive value after digestion. As noted, *in vivo* animal trials to determine the nutritive value and digestibility of feeds are time consuming and expensive and require large quantities of feed. For these reasons *in vitro* systems have been developed and are very useful for the screening of novel animal feed ingredients and additives.

The basis of all *in vitro* rumen screening methods is anaerobic fermentation of a feed sample (substrate) with a buffer solution, simulating ruminant saliva, to produce a filtered rumen liquor which can be analysed. From these *in vitro* tests, it is possible to collect information on fermentation patterns, total gas and methane production, ammonia-nitrogen content, volatile fatty acids, nutrient digestibility and degradability, as well as microbial community composition (Getachew et al., 2004; Smith et al., 2020).

3.1 RUSITEC

The RUSITEC continuous *in vitro* fermentation system (Picture 1) simulates the rumen environment in a standardised, reproducible way, which avoids the inherent variability in individual animals. By combining high-throughput sequencing with a targeted quantitative metabolomics approach, it is able

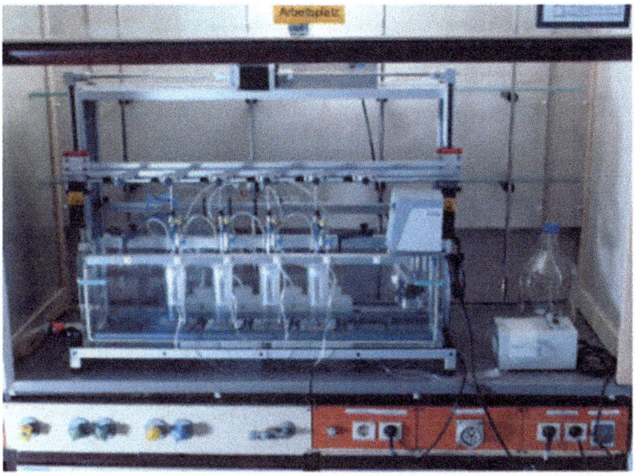

Picture 1 Continuous *in vitro* fermentation system (RUSITEC) or rumen simulation system with six fermentation vessels (Adapted from Birk et al., 2018).

to measure greenhouse gas emissions from feeds, fermentation products and provide a detailed picture of microbiota dynamics (Birk et al., 2018). This method is widely used for studying effects of different diets or feed additives on microbial fermentation patterns, protein synthesis and microbial growth.

The system can maintain ruminal microorganisms indefinitely by feeding them a normal ruminant diet each day and providing the correct physiological conditions in terms of temperature, pH and flow of artificial saliva. To start the incubation, each fermentation vessel is filled under a CO_2 atmosphere with rumen content (liquid from ruminant donors), warm artificial saliva, a nylon bag containing solid rumen digesta (wet weight) and one nylon bag containing the substrate. Fermentation vessels are then submerged in a 39°C water bath. After 24 h, the bag containing the solid rumen digesta is replaced with a new substrate bag. After the following day, the oldest bag is replaced with a new one, so that each bag remains in the vessel for 48 h. During the bag exchange process, vessels are flushed with pure CO_2 to maintain anaerobic conditions. Throughout the experiment, saliva is continuously infused into each vessel. Eight days are allowed to achieve equilibrium before sampling.

Microbial composition, gas production, volatile fatty acid profiles, ruminal pH, among other fermentation variables, are measured from the outflows collected daily in flasks or plastic bags connected to the fermentation vessels. Every day, at the time of substrate bag exchange, collection containers (flasks and gas collection bags) are clamped and replaced with new ones (Birk et al., 2018; Owen et al., 1991; Duarte et al., 2017a).

3.2 In vitro *batch culture*

In this *in vitro* technique, ground particulate substrates, anaerobic media and a rumen fluid inoculum are used to provide information on the fermentation kinetics of ruminant feeds. The technique differs from other *in vitro* procedures as incubations are conducted in gas-tight culture bottles, thus enabling gases to accumulate in the head-space as the fermentation proceeds (Picture 2). A pressure transducer connected to a digital readout voltmeter, together with a gas-tight syringe assembly, is then used to measure and release the accumulated gas pressures from the incubating culture bottles. By repeating this gas-measurement/gas-release procedure at regular intervals during the fermentation, it is possible to construct gas accumulation profiles for feeds by calculating regression-corrected gas volumes (Theodoridou et al., 1994).

3.3 Protein evaluation systems and models

The nutritive value of a new feed ingredient can also be predicted using advanced protein evaluation systems such as the Cornell Net Carbohydrate and Protein System (CNCPS), as well as prediction models of truly digestible nutrient supply to ruminants such as the NRC-2001 model, the DVE/OEB model and the PDI model (Theodoridou and Yu, 2013). The CNCPS is used to characterize the nutrient value of feed ingredients by dividing crude protein and carbohydrate into several subfractions in order to understand degradation characteristics during digestion in ruminants (Fox et al., 2004). Using these

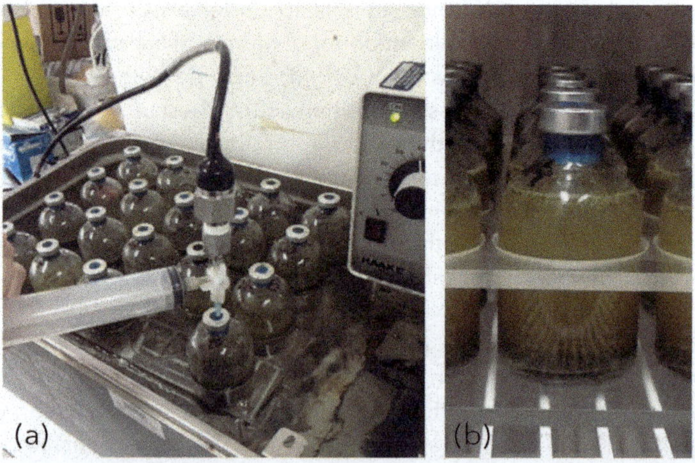

Picture 2 (a) Incubating culture bottles; (b) Culture bottles including the buffer solution, rumen fluid and the studied substrate.

systems and models allows prediction of the different components of a new feed ingredient (i.e. protein, fibre, lipid).

4 Analysing nutrient molecular structure

In order to evaluate the nutritive value of a feed ingredient, most studies focus on chemical composition using traditional 'wet' chemical analysis without any consideration of inherent structural and matrix effects (Yu, 2004). However, the chemical composition of feed (e.g. protein, fibre and carbohydrate content) cannot always explain differences in feed nutrient availability and utilization by animals. It has been found that the quality, utilisation, availability and digestive behaviour of protein are related not only to total chemical composition, but also to the structural make-up of biopolymers in feeds (Picture 3) (Yu, 2004; Aufrère et al., 2013).

The molecular structure of a nutrient, which can change under different processing conditions (e.g. heating), can also play an important role in nutrient availability. Research shows that protein molecular structure is related to the availability of nutrients in the gastrointestinal tract of the animal (Yu and Nuez-Ortín, 2010). Changes in protein, lipid and carbohydrate structure can influence the nutrient value, utilization, availability, and digestive behaviour of a feed. A high percentage of the b-sheet structure may contribute to low access to gastrointestinal digestive enzymes, which results in low protein value and low protein availability (Yu, 2005). Studies of molecular structures are therefore a valuable supplement to traditional feed nutrition studies.

Fourier Transform Infrared Spectroscopy (FTIR) and Synchrotron-Based Fourier Transform Infrared Spectroscopy (SR-FTIR) are potentially valuable tools to obtain information about feed structure simply and rapidly. Figure 1 illustrates a typical animal feed FTIR spectrum showing various functional groups. However, the application of these two spectroscopy techniques in the feed industry is still at an early stage and more research is needed to screen different feeds and feed additives to evaluate their potential in animal nutrition. Figure 2 shows an experimental design to evaluate the potential of a novel animal feed

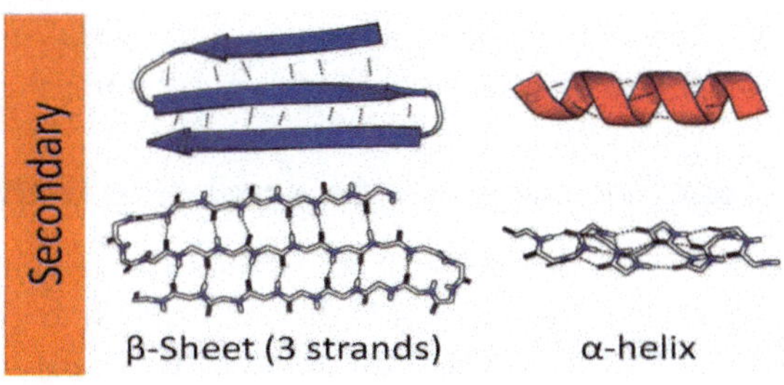

Picture 3 Secondary protein structure of animal feeds.

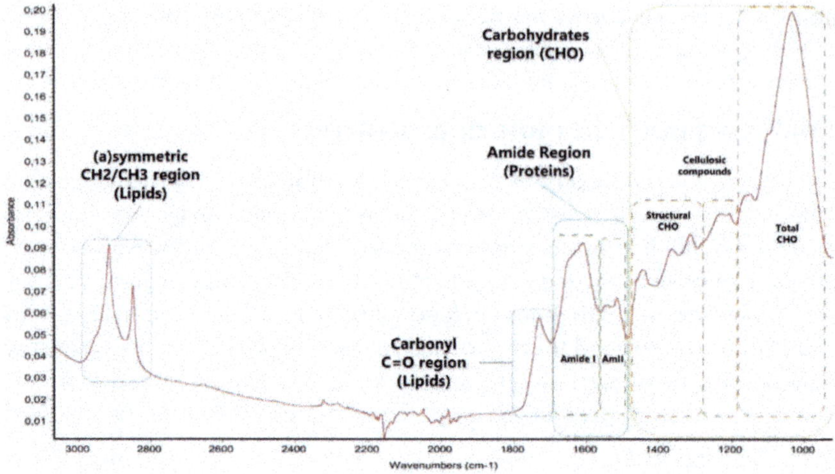

Figure 1 Typical FTIR spectrum of showing various function groups in feed.

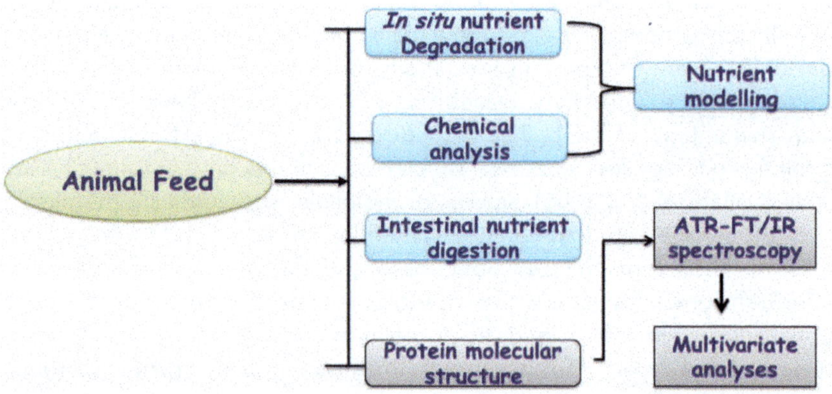

Figure 2 Outline of an experimental design to evaluate the nutritive value of a new animal feed using conventional and spectroscopic techniques.

or additive. The nutritive value can be determined by a combination of lab-based chemical analysis, the use of animal studies and the analysis of nutrient molecular structure. In the long run, the latter technique may take over from traditional methods, such as animal feeding trials and chemical analysis.

5 Analysing bioactive compounds: screening techniques

It has long been common practice to use additives in animal nutrition to: improve gut health and prevent illness (antibiotics, zinc and copper oxide); increase antioxidant content of feed (ethoxyquin, BHA, BHT); or promote animal growth (β-agonists) (Guil-Guerrero et al., 2016). Because of concerns about safety and antibiotic resistance (World Health Organization, 2014), which has led to increasing regulation (e.g. EU Regulation 1831/2003), there has been a trend to search for natural alternatives with similar effects.

The term 'bioactive compounds' generally refers to the plant-derived phytochemicals or secondary metabolites that exert positive effects on animal and human health. Examples include:

- essential oils (EOs) and peptides as an alternative to antibiotics thanks to their antimicrobial properties,
- polyphenols and carotenoids with strong antioxidant properties capable of extending feed shelf life and improving animal health, and
- oligo- and polysaccharides fractions to promote beneficial gut bacteria.

The detection of these compounds in feed materials adds to their potential benefits and value; they can also be isolated and developed as feed additives.

There are also substances of fungal or animal origin with beneficial biological activities in animal nutrition. Some fungi-derived enzymes (phytases, β-glucanases), for example, are often used in feeds to improve feed digestibility (Simon and Igbasan, 2002). Chitin (chitosan) present in mollusc shells and insect exoskeletons has shown antioxidant, antimicrobial and prebiotic potential both *in vitro* and *in vivo* (Sahariah and Másson, 2017). This section focuses on the identification of bioactive chemicals of plant origin since they are the most widely characterised and represent the bulk of potential compounds for use in animal nutrition, especially ruminants.

Figure 3 summarises the main steps in screening raw materials (e.g. seaweeds, agro-industrial by-products, shrubs or food waste) to detect and characterise any bioactive compounds. The first step should include a literature review and evaluation of scientific data to assess likely substances present according to previous studies. Existing information on similar materials provides an idea of what to investigate. Table 1 includes examples of recent studies on agro-industrial and food by-products, seaweeds, shrubs and trees.

5.1 Extraction techniques

The evaluation of any bioactive compound requires a suitable extraction method, which depends on the structure and nature of the raw material and the target chemical group. Extraction is generally preceded by sample pre-treatment. Most extraction methods need the removal of water and reduction of particle size (to increase the surface-to-area ratio) to improve extraction efficiency. A defatting step may also be needed to remove plant waxes or chlorophylls that might interfere with the extraction of compounds of interest such as polyphenols.

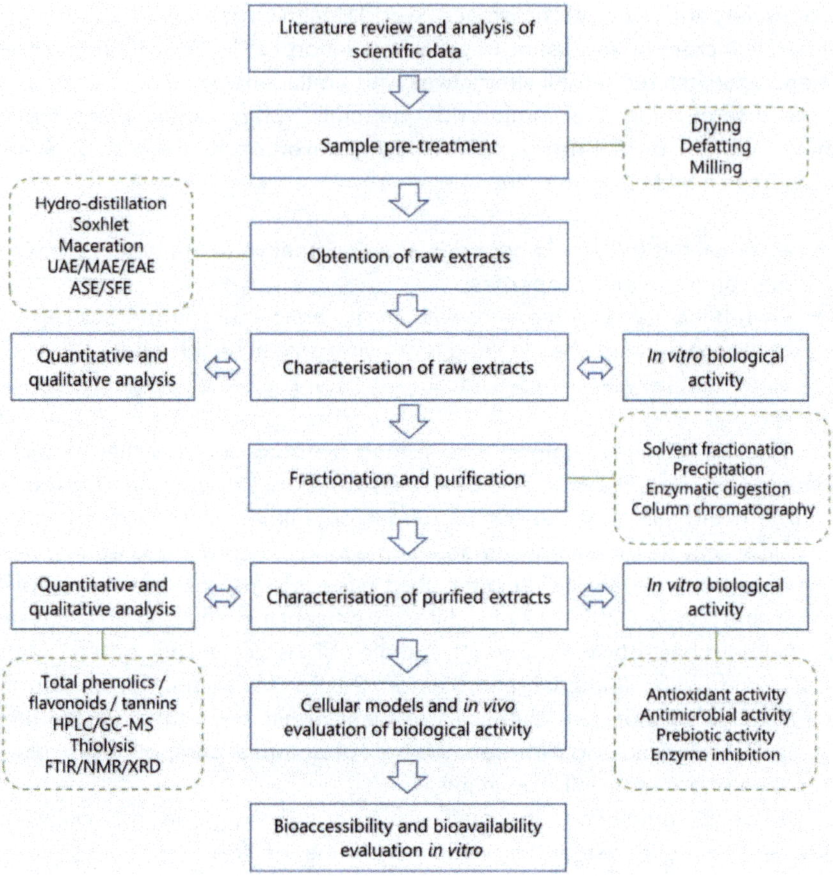

Figure 3 Screening protocol of the bioactive compounds present in raw materials potentially used as feedstuffs.

Hydro-distillation, the Soxhlet system and maceration are the most widespread extraction techniques for screening at laboratory scale. Hydro-distillation is a traditional and still commonly employed method for the extraction of essential oils (Stratakos and Koidis, 2016). The Soxhlet system remains an efficient technique for the extraction of different bioactive compounds (López-Bascón and Luque de Castro, 2019). Maceration involves solvent-liquid extraction by direct interaction of the (dried and milled) raw material with the solvent for a suitable time (from few minutes to several hours) with or without agitation. Despite limitations, these techniques remain popular extraction methods for screening plant bioactive compounds in many laboratories and represent a reference in the evaluation of novel alternatives.

'Green' techniques have been developed to overcome limitations in conventional methods by improving extraction efficiency and yield while reducing extraction time and solvent consumption (Ameer et al., 2017). Some of these methods are similar to maceration in releasing bioactive compounds from cellular compartments: ultrasound (UAE or sonication), microwave (MAE), enzyme (EAE) and pulse-electric field (PEF)-assisted extraction. Other techniques involve modification of the properties of the solvent to improve the extraction process: supercritical fluid extraction (SFE) and accelerated solvent extraction (ASE) or pressurised liquid extraction (PLE). Some are easily applied in most of the laboratories, whereas others require specific and relatively specialised equipment. These techniques are reviewed in Zhang et al. (2018) and Panzella et al. (2020).

For both conventional and 'green' methods, the choice of a suitable solvent or combination of solvents is crucial. Selection typically depends on the polarity of the target compounds. As a reference, the polarity of common solvents from lower to higher is: Hexane < Chloroform < Ethylacetate < Acetone < Methanol < Water. In some cases, there are standard combinations of solvents commonly employed for a specific group of compounds. A mixture of 70:30 acetone:water is, for example, commonly used for bulk extraction of tannins. Extracts with different polarities vary in how well they identify particular fractions (Mgbeahuruike et al., 2018). More detailed information about the selectivity of different solvents can be found in Cowan (1999).

5.2 Fractionation and purification of raw extracts

Raw plant extracts usually combine substances with different biological activities. It may be necessary to perform further separation and purification steps to obtain active fractions or pure compounds, for example, a specific macromolecular-rich crude fraction. Whilst a crude fat fraction can be easily obtained using a non-polar solvent (hexane/petroleum ether), protein and carbohydrate (polysaccharides) require different precipitation and isolation methods (Habeebullah et al., 2020; Hamed et al., 2020).

The purification of specific phytochemical fractions is generally achieved by column chromatographic techniques, based on the adsorption properties of the solid phase and the polarity of the liquid phase. Sephadex LH-20, for example, is commonly employed for the fractionation and purification of tannin extracts from other phenolic compounds (Feng et al., 2014). Thin-layer chromatography (TLC), preparative liquid chromatography and membrane filtration are other separation techniques employed. Zhang et al. (2018) provide a list of available standard techniques.

5.3 Qualitative and quantitative characterisation

Bioactive compounds present in raw extracts can be quantified either in bulk (total content of specific group) or by individual compounds after

Table 1 Examples of bioactive compounds and biological activity screening in raw materials potentially used as feedstuffs

Plant material	Pre-treatment	Extraction method	Bioactive compound	Characterisation	Biological activity	Reference
AGRO-INDUSTRIAL BY-PRODUCTS						
Hazelnut involucre	Air-drying and milling	Maceration – 50–100% acetone	Polyphenols and condensed tannins	LC/MS TPC TFC TT	Antioxidant activity (ABTS/DPPH/FRAP) Enzyme inhibitory activity (tyrosinaes/α-glucosidase)	Rusu et al., 2019
Bamboo shoots culm	Defatting and decolouring, air-drying and milling	ASE – Water	Polysaccharides	CHO composition FT-IR spectroscopy Thermal analysis	Rheological properties (viscosity/frequency sweep) Antioxidant activity (ORAC/DPPH/reducing power/Fe^{2+} chelation)	Chen et al., 2018a
Avocado peel and seed	Freeze-drying and milling	UAE – 80% ethanol	Phenolic compounds	TPC HPLC-DAD/MS	Antioxidant activity (DPPH/FRAP/ABTS/deactivation of ROS) Cytotoxicity (MTT assay/TNF-α production/NO production)	Tremocoldi et al., 2018
Wheat co-products	None	*In vitro* enzymatic digestion	Fibrous compounds	Proximal composition	Prebiotic activity (gas production, SCFA)	Jha et al., 2012
Artichoke floral stems	Forced-air drying and milling	Maceration – Butanol/Ethyl acetate/Ethanol/Methanol	Polyphenol compounds (phenolic acids, flavonoids and diterpenes)	TPC TFC TT HPLC-DAD-MS	Antioxidant activity (DPPH/ABTS/FRAP), antimicrobial activity (Disk diffusion/MIC/MBC) Anti-Ache and anti-denaturing protein activity	Mejri et al., 2020
Pistachio green hulls	Forced-air drying, milling and defatting	Maceration – water (+ purification)	Polysaccharides	^{13}C NMR FTIR UV/vis X-ray diffraction TLC	Water activity (Aw) Antioxidant activity (DPPH/ABTS/reducing power/β-carotene bleaching inhibition)	Hamed et al., 2020

Asparagus by-products	Fresh	Maceration – 80% ethanol	Flavonoids Saponins	HPLC-DAD/MS	Antioxidant activity (DPPH/Ferric reducing power/linoleic acid oxidation model/p-anisidine)	Fuentes-Alventosa et al., 2013
Broccoli stems and leaves	Freeze-drying and milling	Maceration – ethanol/hexane/methanol	Glucosynolate Vitamin E/K Carotenoids Phenolic comps	HPLC-DAD TPC	Antioxidant activity (DPPH)	Liu et al., 2018
Tomato seeds and peels	Air-drying and milled	UAE – methanol:ethyl acetate:petroleum ether/80% methanol	Carotenoids/Phenolic compounds	HPLC-DAD/MS	Antimicrobial activity (MIC) Antioxidant activity (DPPH)	Szabo et al., 2019
Olive cake and leaves	Air-drying and milling	Maceration – 80% Methanol	Polyphenolic compounds	TPC TFC HPLC-QTOF/MS	Antioxidant activity (DPPH/Ferrous iron chelation/ferric reducing power) Antimicrobial activity (Disk diffusion/MIC/MBC)	Moudache et al., 2020
Peanut shells	Air-drying and milling	MAE + EAE (cellulase – 70% ethanol) (+ resin purification)	Polyphenolic compounds	TPC	Antioxidant activity (DPPH/OH· scavenging assay/reducing power/β-Carotene bleaching assay) Antibacterial activity (disk diffusion/Oxford cup/MIC)	Zhang et al., 2013
Pomegranate peel	Freeze-drying and milling	*In vitro* enzymatic digestibility	Phenolic compounds Non-digested fraction	TPC TFC HPLC-DAD	Antioxidant activity (DPPH/ORAC/ABTS/FRAP/Fe^{2+} chelation) Prebiotic activity (SCFAs)	Gullón et al., 2015
Grape cane	Forced-air drying and milling	UAE – 50% Acetone Water:ethanol:hexane	Phenolic compounds Carotenoids	TPC HPLC-DAD	Antioxidant activity (ORAC/DPPH/ABTS)	Ferreyra et al., 2020
Guava seeds	Drying	Soxhlet – hexane	Phenolic compounds	HPLC-DAD-MS	Antioxidant activity (DPPH/DCF-fluorometric)	Prommaban et al., 2020

(Continued)

Table 1 (*Continued*)

Plant material	Pre-treatment	Extraction method	Bioactive compound	Characterisation	Biological activity	Reference
TREES AND SHRUBS						
Chestnut leaves	Freeze-drying and milling	Maceration - 60% Ethanol	Polyphenolic compounds	HPLC-DAD TPC O-diphenols (Arnow assay) TFC TT	Antioxidant activity (FRAP/DPPH)	Vella et al., 2019
Rosemary leaves	Forced-air drying	Hydro-distillation	Essential oils	GC-MS	Antimicrobial activity (disk diffusion/MIC/MBC)	Jordán et al., 2013
	Forced-air drying, milling and defatting	Soxhlet - 100% methanol	Polyphenol compounds (phenolic acids, flavonoids and diterpenes)	HPLC-UV/vis	Antioxidant activity (DPPH/ABTS/FRAP)	
Guinea pepper seeds and leaves	Air-drying and milling	Sequential maceration/ fractionation - hexane/ chloroform/ethanol/ methanol	Alkaloids	HPLC/DAD HPLC/Q-TOF-MS	Antimicrobial activity (disk diffusion/MIC/MBC)	Mgbeahuruike et al., 2018
Buckthorn (*Ziziphus* spp.) leaves	Drying and milling	Maceration - 100% methanol	Phenolic compounds	TPC TFC TT LC/MS	Antioxidant activity (DPPH/TAC) Antimicrobial activity (disk diffusion)	Yahia et al., 2020
***Ficus altissima* leaves**	Freeze-drying and milling	UAE - 100% methanol (+ HPLC-purification)	Condensed tannins	TPC Thiolysis-HPLC/MS	Antioxidant activity (DPPH, DNA protective effect) Tyrosinase inhibitory assay	Deng et al., 2016

Lotus pedunculatus* and *Trifolium repens	Freeze-drying and milling	Maceration - 70% Acetone (+ Sephadex LH-20 purification)	Condensed tannins	$^{1}H/^{13}C$ NMR Thiolysis-HPLC/DAD	Protein precipitation assay (RDA)	Zeller et al., 2015
Flamboyant tree (*Delonix regia*) leaves	Freeze-drying and milling	Maceration - 70% Acetone (+purification by sephadex LH-20)	Phenolic compounds	Thiolysis-HPLC/MS TPC CT	Antioxidant activity (ABTS/DPPH/CUPRAC) Tyrosinase inhibitory activity (fluorescence/Cu^{2+} chelating activity)	Feng et al., 2014
Sainfoin (*Onobrychis viciifolia*)	Freeze-drying and milling	Maceration - 70% Acetone	Condensed tannins	Thiolysis-HPLC (direct analysis)	*In vitro* rumen fermentation (pH, VFA, NH3-N) Protein precipitation (RDA assay) (CT extract)	Theodoridou et al., 2010
SEAWEEDS						
Brown seaweeds (seven species)	Freeze-drying and milling	EAE - acetate/phosphate buffer (± fractionation)	Phenolic compounds Proteins Polysaccharides	TPC Bradford DNS method	Antioxidant activity (DPPH/Fe chelation/reducing power/Lipid peroxidation inhibition) Antimicrobial activity (Disk diffusion)	Habeebullah et al., 2020
Ascophyllum nodosum* and *Fucus serratus	Freeze-drying and milling	Maceration - 70% Acetone (± SPE fractionation)	Phlorotannins	Folin-Ciocalteu $^{1}H/^{13}C$ NMR	MIC/MBC Cell viability assay	Ford et al., 2020
Fucus serratus* and *Polysiphonia fucoides	Freeze-drying and milling	Maceration - 50/100% Ethanol	Phenolic compounds	TPC HPLC-DAD	Antioxidant activity (DPPH/reducing power/Fe^{2+} chelating activity) Lipid oxidation in oil-in-water emulsion (peroxide value/volatile organic compounds)	Habeebullah and Jacobsen, 2015
Ecklonia radiata	Oven-drying and milling	EAE-extraction and fractionation	Polysaccharides		Prebiotic activity (SCFA and microbial enumeration - PCR)	Charoensiddhi et al., 2016

suitable separation. Different techniques have been developed for the bulk quantification of the main bioactive groups: total phenolic compounds using the Folin-Ciocalteau method (Singleton and Rossi, 1965); flavonoids (Zhishen et al., 1999); condensed tannins using HCl:butanol (Grabber and Zeller, 2020; Vanillin, Julkunen-tiitto, 1985); and carotenoids (Lichtenthaler, 1987). Most of these methods are based on the reaction of a specific chemical feature of the target molecule with a reagent, yielding a colourimetric compound that can be determined spectrophotometrically. These bulk quantification techniques tend to be unspecific as many other similar compounds or impurities can affect the results. However, they are quick, cheap and easily implemented in most laboratories and are therefore useful for the screening of novel raw materials.

For the identification and quantification of individual substances present in raw extracts, chromatographic techniques are often employed. Gas chromatography (GC) is typically used for the analysis of volatile terpenoid substances found in essential oils. Bioactive substances in raw extracts are separated by (high- or ultra-high pressure-)liquid chromatography (LC/HPLC/UPLC), coupled to UV/vis or diode-array (DAD) detectors when the composition is known and reference standards are available. Mass spectrometers are used to identify unknown compounds or when additional qualitative information is required. Indeed, mass spectrometry (MS) is considered one of the most powerful tools in the analysis and identification of bioactive compounds. These chromatographic methods have shown to quantify phenolic compounds (Mejri et al., 2020), condensed tannins (Deng et al., 2016), carotenoids (Szabo et al., 2019) and tocopherols (Liu et al., 2018) (see Table 1).

Novel fast and non-destructive techniques used for identifying macronutrients can also be used to characterise bioactive compounds in feedstuffs with minimal pre-processing. Fourier-transformed infrared (FTIR) spectroscopy is a valuable tool for the quantification of bioactive compounds (Lee et al., 2015) and biological activity (Deyrieux et al., 2018), as well as for the qualitative determination of structural features, identification of organic functional groups and molecular 'fingerprints' of compounds (Hamed et al., 2020). Nuclear magnetic resonance (NMR) imaging can be used to obtain both quantitative (Campbell et al., 2020) and qualitative (Ford et al., 2020) information about bioactive compounds in raw extracts.

The benefits of consumption of condensed tannins (CTs) by ruminants are related to their protein binding ability, including reducing bloating, antiparasitic properties, improvement of nitrogen utilisation and reduction of greenhouse gas emissions. Structural characterisation of CTs requires information about molecular size (or mean degree of polymerisation - mDP), prodelphidines:procyanidin (PD:PC) and *cis:trans* ratios to properly understand the biological activity of CTs in relation to ruminant nutrition.

However, the analysis of CTs is sometimes challenging because of significant structural diversity derived from flavan-3-ol subunit combinations and linkages. Techniques require acid cleavage of the interflavan links, releasing the individual flavan-3-ol subunits, which are captured by benzyl-mercaptan (thiolysis) (Gea et al., 2011) or phloroglucinol (Kennedy and Jones, 2001).

Acid-cleavage methods for characterisation of CT structure are challenging. A quick and practical method to assess the biological activity of CTs is the radial diffusion assay (RDA). This quantifies the ability of CTs to precipitate a reference protein (bovine serum albumin) embedded in an agar solution by measuring the diameter of the opaque complex formed after the diffusion of the tannin solution (Makkar, 2000). This method also gives an indirect estimation of the content. The PEG-binding assay is another method for *in situ* determination of tannins in plant samples without extraction (Silanikove et al., 1996). Zeller (2019) provides a detailed review of methods for the analysis of CTs.

6 Analysing bioactive compounds: *in vitro* evaluation of antioxidant, antimicrobial, prebiotic and enzymatic activity

In vitro modelling is useful for screening biological activity as it is relatively easy and does not require sophisticated handling techniques.

6.1 Antioxidant activity

Antioxidant compounds were initially added to feeds to prevent lipid oxidation, thus extending shelf life. However, conditions affecting livestock such as stress, ingestion of toxic substances (e.g. pesticide residues, mycotoxins or heavy metals), nutritional deficiencies or illness can increase the formation of free radicals - unstable chemical species capable of reacting with lipids, DNA or proteins - resulting in potential oxidative and cellular damage (Fellenberg and Speisky, 2006). Improving antioxidant status through diet has been demonstrated to be an effective strategy to prevent such damage, either by direct scavenging activity or indirectly by activation of biological antioxidant pathways (e.g. enzymes, vitamins) (Ortuño et al., 2015).

Effective *in vitro* characterisation of antioxidant activity requires the use of complementary techniques covering different antioxidant mechanisms of target compounds. Detailed information on specific techniques can be found in Shahidi and Zhong (2015) and Craft et al. (2012).

Radical scavenging methods: these techniques are generally classified according to the chemical reaction involved in antioxidant activity:

- hydrogen-atom transfer (HAT), and
- single-electron transfer (SET).

Among the most common HAT methods employed are:

- Oxygen Radical Absorbance Capacity (ORAC) (Prior et al., 2003),
- chemiluminescence (Alho and Leinonen, 1999),
- Total Radical Trapping Antioxidant Parameter (TRAP) assay (Lussignoli et al., 1999), and
- crocin/β-carotene bleaching assay (Kampa et al., 2002).

Routine SET methods are:

- 2,2-Diphenyl-1-picrylhydrazyl (DPPH) radical scavenging assay (Brand-Williams et al., 1995).
- Trolox equivalent antioxidant capacity assay (also known as the ABTS assay due to the target radical employed in the reaction((Re et al., 1999).

These methods have been criticised for not fully correlating with *in vivo* capacity, as they are based on the scavenging capacity of bioactive substances against specific types of radical species. Nevertheless, they provide useful information about intrinsic antioxidant potential. They are therefore widely used for screening the antioxidant potential of feedstuffs.

Non-radical redox potential-based methods: antioxidant efficacy can be measured by the reduction capacity of bioactive compounds, based on a SET mechanism. The main assays involve measurement of ferric (FRAP; Pulido et al., 2000) and cupric (CUPRAC; Apak et al., 2004) reducing antioxidant capacity.

Metal chelation capacity: the interaction of transition metal ions with susceptible oxidative molecules (PUFA) can promote lipid oxidation through the Fenton reaction and decomposition of hydroperoxides. The ability of bioactive compounds to form complexes with metal ions is another indicator of their antioxidant capacity. Divalent iron and copper ions are used to measure antioxidant metal chelation capacity. Standard tests are the ferrozine assay (Dinis et al., 1994) and the tetramethylmurexide (TMM) assay (Shimada et al., 1992).

Lipid oxidation measurement: the formation of radicals from highly-oxidative (poly-)unsaturated fatty acids is generally considered the triggering mechanism for the formation of hydroperoxides and conjugated dienes (primary oxidation). These then attack biological substances (DNA, proteins) or decompose into secondary oxidation compounds (hydrocarbons, alcohols, ketones, aldehydes) which reduce feed quality. One way of assessing the *in vitro* antioxidant capacity of bioactive compounds is quantification of oxidation inhibition using lipid models. Model systems include bulk oils, water-in-oil emulsions, triacylglycerols or individual free PUFA (i.e. linoleic acid). Changes in the content of unsaturated fatty acids, the formation of primary or secondary oxidation compounds and oxygen consumption can be quantified at different points in time and accelerated or not by the presence of radical initiators or heating (Craft et al., 2012). Common methods include measurement of:

- hydroperoxides (Peroxide value – PV) (Deyrieux et al., 2018; Shantha and Decker, 1994),
- conjugated dienes (CD) (Farmer and Sutton, 1943),
- secondary oxidation compounds (p-Anisidine value – pAV),
- thiobarbituric acid-reactive substances (TBARS) (Botsoglou et al., 1994), and
- volatile organic compounds (Ortuño et al., 2016).

The total oxidation ('Totox') value has been used for primary (PV) and secondary (pAV) oxidation assessment (Totox value = pAV + 2PV) (Laguerre et al., 2007). Official methods and protocols for the measurement of lipid oxidation can be found in the American Oil Chemistry Society manual (www.aocs.org).

6.2 Antimicrobial activity

The reduction in the use of antibiotics to prevent zoonotic antibiotic resistance has encouraged the search for suitable natural alternatives (WHO, 2014). This requires evaluation of antimicrobial activity of candidate compounds (Murphy et al., 2017). This section reviews common methods to screen raw extracts, fractions or pure compounds in raw materials intended to be used as feed ingredients or additives. Choosing the most suitable method depends on its particular advantages and limitations. A combination of methods is usually recommended to fully evaluate the antimicrobial activity of the target compounds.

Agar disk-diffusion method: this is a routine method used in many clinical microbiology laboratories for testing antimicrobial susceptibility. Approved standard procedures are published by the Clinical and Laboratory Standards Institute (CLSI) for bacteria and yeast testing (CLSI, 2015). The method involves growth inhibition of a specific microorganism inoculated in agar plates by the target compound, which is incorporated onto filter paper discs at the desired concentration. The target substance diffuses into the agar during incubation, and the measurement of the diameter of inhibition determines its activity. The main drawbacks of this method are the inability to differentiate between bactericidal and bacteriostatic effects since bacterial growth inhibition does not necessarily mean bacterial death, and the inability to determine the minimum inhibitory concentration (MIC), as it is not possible to quantify the amount diffused into the growth medium. However, since it is simple and cost-effective, it is ideal for screening purposes.

Agar well diffusion method: this method is similar to the above. Instead of a filter paper disc, a specific volume of the tested bioactive compound is poured in a well (6–8 mm of diameter) of inoculated agar. Antimicrobial activity is assessed by measuring the size of the inhibitory growth diameter (Valgas et al., 2007). There is a variation of this method, called the agar plug diffusion method, designed to study competition between microorganisms (Elleuch et al., 2010). The two targeted microorganisms are incubated in different agars, and, subsequently, a cylindrical piece of one agar is aseptically removed and substituted by a similar piece of the other microorganism. The inhibitory potential of the molecules secreted by one of the microorganisms (bacteriocins) on the other can then be evaluated.

Broth/agar dilution methods: broth and agar dilution methods represent the most appropriate way to determine minimum inhibitory concentration (MIC) and minimum bactericidal concentration (MBC). The MIC value is defined as the lowest concentration (mg/mL or %) of the tested compound that inhibits the visible growth of the target microorganism. MBC is the lowest concentration that can inactivate 99.9% of bacteria from the initial population. The determination of the MBC value requires the calculation of bacterial levels (CFU/ml) by culture dilutions and subcultures. The main standards for antimicrobial susceptibility testing using this method are from CLSI and the European Committee on Antimicrobial Susceptibility Testing (EUCAST). These guidelines provide uniform and practical procedures for testing which allows comparison of results between different laboratories (Pfaller et al., 2004). The broth dilution method, commonly used in animal feed testing, consists of visual or spectrophotometric assessment of the bacterial growth in two-fold dilutions of the tested compound in a liquid growth medium after a defined time (16–24 h) (Wiegand et al., 2008). Microdilution techniques offer a quicker, safer and more reproducible alternative if the required equipment (automatic plate reader) is available (CLSI, 2015).

Time-Kill assay (time-kill curve): this method is similar to broth dilution assays, although periodic measurements are made to study bactericidal or bacteriostatic activity of the tested compound over a defined time (24-48 h). In comparison with the previous methods, the main advantage is the possibility of monitoring the effect of different concentrations of the active substance on the growth stage of the target microorganism (Pfaller et al., 2004).

Flow cytometry: by using fluorescence probes that bind to microbial cells, flow cytometry measures the light scattering of cells in suspension and the fluorescence of cellular contents or metabolic activity (Álvarez-Barrientos et al., 2000). The antimicrobial activity of the target compounds can be described in terms of changes in cell shape and fluorescence (Álvarez-Ordóñez et al., 2013). One of the most important advantages is the possibility of detecting microbial subpopulations less susceptible to active substances. Flow cytometry and fluorescence probes allow quantitative, accurate and rapid susceptibility tests of bacterial response to a panel of antimicrobial agents before determining the most suitable one for a specific application. The widespread use of this method is limited by the cost and availability of flow cytometry equipment in laboratories.

6.3 Prebiotic activity

Testing animal feed does not only involve testing the ability to inhibit (bacteriostatic) or reduce (bactericidal) numbers of undesirable bacteria (generally pathogenic species) but also the promotion of beneficial bacteria. This topic is gaining interest due to the research done initially in human nutrition. 'Prebiotics' are compounds resistant to gastric and pancreatic enzyme digestion but fermentable by colonic and caecal bacteria. The promotion of 'beneficial' bacteria (e.g. *Lactobacillus*, *Bifidobacterium*) by prebiotic substances increases the production of bacteriostatic substances (bacteriocins, surfactant) and other metabolites (organic acids, short-chain fatty acids). These compounds can limit the growth of pathogenic bacteria (*Salmonella, Clostridia, E. coli*), improve gut health and exert other benefits (e.g. in boosting immune system function) (Aquino et al., 2017). Different agro-industrial by-products and co-products, fibrous feedstuffs and seaweeds contain compounds with prebiotic potential.

As discussed previously, *in vitro* models have been used to study the prebiotic potential of new animal feeds for ruminants. Studies of the structure and activities of the gut microbiota, functional interactions between gut microbes, and relationships between microbes and host cells, are important topics for future prebiotic research in ruminants. 'Meta-omic' approaches (metagenomic, metatranscriptomic, metaproteomic and meta-metabolomic analyses) are powerful tools for analysing the relationships between the gastrointestinal tract, microbial community and host metabolism (Uyeno et al., 2015). *In vitro* methods for evaluation of the mechanisms involved in the promotion of GI microflora in monogastrics are similar to those employed in human nutrition and, therefore, can easily be adapted with minor modifications to assess animal feed ingredients.

Supplemented medium growth: one of the simpler and quicker methods consists of supplementation of a culture medium (generally Man, Rogosa and Sharpe - MRS) with the lyophilised feedstuff or extracted prebiotic substance. An isolated or mixed group of bacteria (*Lactobacillus* or *Bifidobacterium*) are then inoculated in the agar or broth, and the prebiotic effect assessed by bacterial growth, changes in the pH and/or the quantification of fermentation products (SCFAs) at the end of the incubation period or at different time points. Results of the tested compound are contrasted with the use of positive (inulin, fructooligosaccharides) and negative (non-carbon source) controls. This method has been successfully applied for the prebiotic evaluation of different agro-industrial by-products (Duarte et al., 2017b; Gómez et al., 2014).

Anaerobic batch culture: in a similar way to the previous method, it is possible to assess the prebiotic activity of a feedstuff or specific compound by incubation with a bacterial inoculum (Gullón et al., 2015; Jha et al., 2011). In this case, the sample is incubated with the tested inoculum in a pH (buffer), anaerobic (CO_2), and temperature (37-39°C) controlled environment for 24+ hours. The microbial inoculum generally consists of a known mixture of bacteria or, if possible, a faecal sample of the target animal (Bindelle et al., 2007). The prebiotic activity can be assessed in a similar way (microbial quantification, pH changes, SCFA formation), as well as by the quantification of the gas produced (Theodoridou et al., 1994). Recently, molecular techniques (polymerase chain replication - PCR; fluorescence in situ hybridisation - FISH) have been developed for bacteria identification and quantification, which avoids the loss of uncultivable organisms, reduces time and costs associated with plate agar counts and can be species-specific. These methods are spreading quickly among laboratories due to their efficacy and increasing affordability.

The sample containing the bioactive compound of interest can be evaluated in different ways: the whole feedstuff, after isolation and purification of the fraction or substances of interest, or as a residue after *in vitro* enzymatic digestion (Jha et al., 2011). The latter requires pre-incubation of the feedstuff with gastrointestinal enzymes (Boisen and Fernández, 1997) and further collection, drying and homogenisation of the non-digested material. In this way, it is possible to mimic fermentation of the fraction in the large intestine of the animal after gastric and small intestine digestion.

6.4 Enzyme inhibitory activity

Another biological activity of phytochemicals is their interaction with enzymes. This is another active field of research in human therapeutics which researchers in the animal science sector can also take advantage of. Different phytochemicals have shown an inhibitory effect against angiotensin-converting enzyme (ACE), cholinesterases, tyrosinases and cyclooxygenases, which are related to inflammatory and degenerative diseases (Gonçalves and Romano, 2017). Some of these effects could help to improve animal health and welfare.

However, whilst such activity can be beneficial for humans, it can be deleterious for animals. For example, the inhibitory activity of phytochemicals on some digestive enzymes (α-amylase, pancreatic lipase), which can ameliorate diabetes or obesity in humans, may negatively affect digestibility and performance in food-producing animals (Rohn et al., 2002).

7 Case study: seaweed as animal feed

Seaweeds have long been employed as a source of livestock fodder in coastal regions (Evans and Critchley, 2014). However, their use declined due to the shift to more intensive methods of production as well as low palatability and the presence of anti-nutritional factors. There has been a renewed interest in inclusion of seaweed as an animal feed ingredient due to the need to reduce pressure on land use, the need for more sustainable and environmentally-friendly sources of feed and the potential to reduce ruminant methane emissions (Maia et al., 2016; Kinley and Fredeen, 2016). Research suggests that seaweed components such as chelated minerals, complex carbohydrates, polyunsaturated fatty acids, protein and bioactive compounds may exert beneficial effects (Kumar et al., 2008).

On the northern coasts of Ireland and Britain brown seaweeds (*Phaeophyceae*) are an abundant natural resource with potential as an alternative animal feed. Four of the main brown seaweed species from two families in this area have been studied for their suitability as ruminant fodder: Laminariales (*Laminaria digitate* and *Saccharina latissima*) and Fucales *(Ascophyllum nodosum* and *Fucus vesiculosus*) (Picture 4).

Seasonal variability is a key factor affecting the composition of seaweed. It is important to specify species, season and geographical location when assessing the quality of seaweeds as livestock feed. Initial characterisation focused on the chemical composition and *in vitro* dry matter digestibility (IVDMD) of the four types of seaweeds during different seasons. Results showed high inter- and intraspecies variation in crude protein (CP), fibre (NDF) and crude ash content of the different species throughout the year (Table 2). Differences in chemical composition between species were reflected in ruminal IVDMD, evaluated using the Daisy^{II} incubation method (Holden, 1999), which showed that Laminariales had higher digestibility than Fucales.

Mineral content is high in both Laminariales (25% DM) and Fucales (≈20% DM). The mineral profile of the four species was evaluated by X-ray fluorescence (XRF), a fast and straightforward method for the quantification of macronutrients (Na, S, K, Cl, K and Ca) and trace elements (I, As, Pb). High Na, Cl and K and iodine content suggest the possible use of brown seaweeds as mineral supplements for livestock. However, it is important to avoid the risk of overdose, especially in the case of iodine, as well as risks of contamination

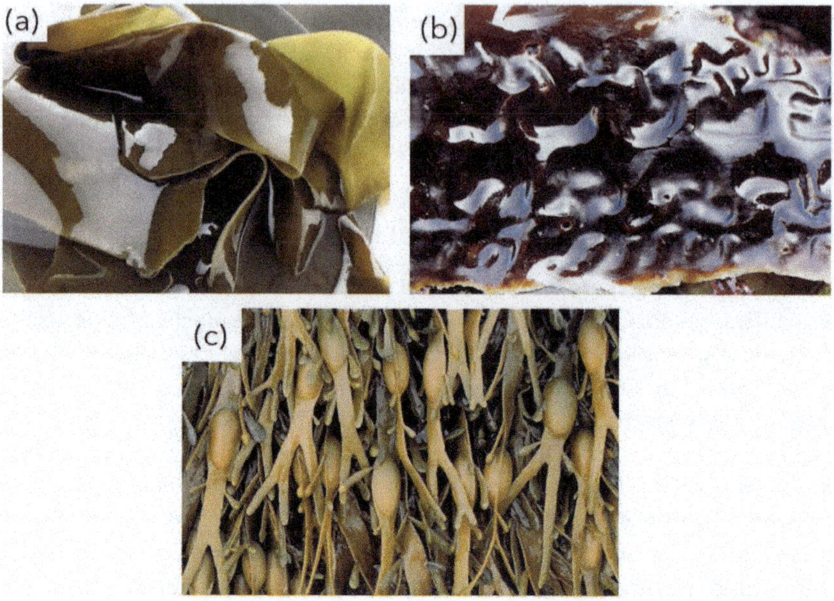

Picture 4 Different seaweed species (a) *Laminaria digitate,* (b) *Saccharina latissimi,* (c) *Ascophyllum nodosum*.

from heavy metals such as arsenic and lead (check Directive 2002/32/EC). By assessing chemical variation due to seasonal and other factors, the use of rapid screening analysis techniques such as XRF help to decide when to harvest seaweed and to ensure it meets animal requirements. Calibration methods based on infrared spectroscopy (NIR and MIR) in combination with chemometrics analysis have been used to evaluate nutritive value and rumen digestibility. Results showed accurate prediction ($R^2P>0.95$) of CP, fibre and polyphenol content with NIR spectroscopy.

Recent research has identified phlorotannins (a polyphenol) in brown seaweeds which possess health-promoting properties, antimicrobial activity and may be involved in the reduction of methane emissions through the alteration of rumen fermentation (Evans and Critchley, 2014). Total phenolic content (TPC) was screened throughout the year in the four different seaweeds species. 70% acetone in water was found to be the most suitable solvent mixture to obtain bulk extracts using the colourimetric Folin-Ciocalteau method (Ford et al., 2019). Results showed that Fucales (*A. nodosum* and *F. vesiculosus*) contained up to 6 times higher TPC than Laminariales (*L. digitata* and *S. latissima*), in agreement with previous studies (Connan et al., 2004). Seasonal intraspecies variation was also found (Table 2) with different peaks depending on the family analysed: late summer-autumn peaks for Laminariales and winter-summer peaks for Fucales.

Table 2 Seasonal variation of two seaweed orders a) Laminariales and b) Fucales

	Laminariales				Fucales			
(%DM)	Winter	Spring	Summer	Autumn	Winter	Spring	Summer	Autumn
DM	12.7	14.7	22.6	18.7	21.3	21.6	23.3	24.1
CP	15.2	13.7	5.42	7.80	11.1	9.52	4.67	6.74
Ash	31.5	28.7	16.9	21.8	22.2	22.6	13.5	18.1
NDF	63.2	46.9	47.1	51.3	54.2	43.5	43.0	45.7
IVDMD	94.5	79.6	86.1	94.4	58.7	63.2	56.8	59.3

The characterisation of seaweed tannins is challenging due to different sizes and types of subunit linkages. Nuclear magnetic resonance (NMR) spectroscopy has been used to improve quantitative and qualitative information on phlorotannins. The intra- and inter-differences in TPC throughout the seasons were analysed by quantitative ^{1}H-NMR (qNMR) in the whole seaweed biomass (Ford et al., 2020). ^{13}C-NMR spectra were also employed to calculate the difference in linkages of phenolics between fresh and silage seaweed species from extracts prepared using solid-phase extraction (SPE) to remove sugar and other non-phenolic compounds (Campbell et al., 2020). The stability of phenolic compounds was also observed by investigating the molecular structure of the compounds before and after ensiling using nuclear magnetic resonance (NMR) (Fig. 4). Quantitative results differed between colourimetric and qNMR methods, reflecting the challenges in determining complex polyphenols in biological material. Structural differences which reflect the major types of linkages (C–C vs C–O–C) between species can be detected by ^{13}C-NMR; these differences may be related to the biological activity of the extracted polyphenols. Finally, different biological activities of interest related to the presence of phlorotannins have also been evaluated.

It has been suggested that phlorotannins in brown seaweeds may reduce ruminal methane production (Wang and McAllister, 2011). A preliminary *in vitro* study was performed to assess the inhibitory methanogenic effect of the four seaweed species, using the method described by Theodoridou et al. (1994) with lucerne as a negative control. Fucales (*A. nodosum* and *F. vesiculosus*) showed reduced total gas and methane production in comparison

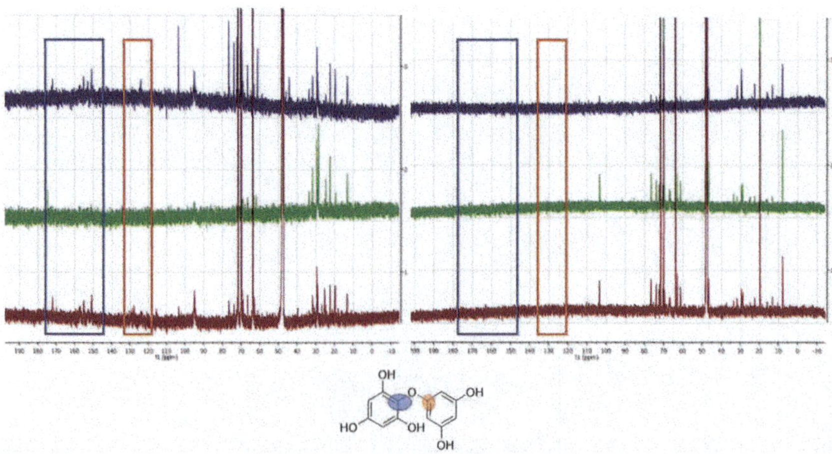

Figure 4 13C-NMR seaweed spectra of FV and SL. The coloured spectra show the seaweeds before (FV0, SL0: blue) and after ensiling, with (FVi, SLi: green) and without (FVu, SLu: red) the inoculant. The blue and orange boxes represent the C-O-C bonds shown on the chemical structure above (Campbell et al., 2020).

with Laminariales (*L. digitate* and *S. lattissima*), corresponding with higher phlorotannin content. Nevertheless, all seaweed species were effective in inhibiting methane production when compared with lucerne. Results showed a promising anti-methanogenic effect of seaweeds, either included as an ingredient or polyphenol extract in the diets of ruminants (Fig. 5).

Polyphenolic compounds have traditionally attracted interest for their antimicrobial properties. Seaweed polyphenolic extracts were evaluated to obtain the minimum inhibitory concentration (MIC) and minimum bactericidal concentration (MBC), as well as to determine cell membrane permeability and intracellular ATP levels of different pathogenic bacteria (*E. coli, Salmonella* spp. *and Streptococcus suis*). *A. nodosum* extracts showed the most significant antimicrobial potential, related to their chemical structure detected by NMR. However, a study by our team showed that the high TPC limited the ensilability of *Fucus vesiculosus* in comparison with *Saccharina latissimi* (Campbell et al., 2020).

The overall results of chemical characterisation show that these four species of seaweed could be used as mineral supplements for livestock. If seasonal variations are closely monitored. Thanks to rapid screening methods such as NIRS, it may be possible to use brown seaweeds as a full ingredient, directly or after ensilage, when CP content and digestibility are high, as may be the case for Laminariales during the winter-spring season. The presence of phlorotannins offers added value through their antimethanogenic potential and antimicrobial properties.

8 Summary and future trends

Brown seaweeds are a promising animal feed due to a suitable amino acid profile, trace elements, vitamins and bioactive compounds which have

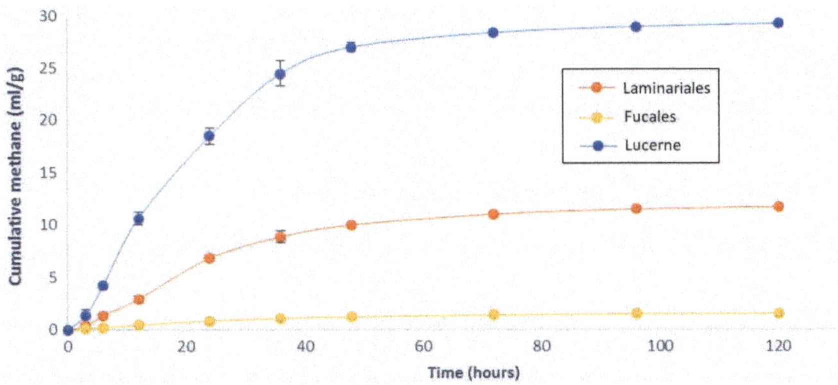

Figure 5 *In vitro* antimethanogenic effect of seaweed in comparison with a forage (lucerne).

antioxidant and antimicrobial activities. Seaweeds can also contribute to freeing up land to grow crops for human consumption. They can also add value to ruminant production systems by reducing greenhouse gas emissions and improving animal productivity and the quality of animal products. Seaweed supplements in livestock increase iodine content in animal products, which has been found to be a good way to provide iodine to deficient human populations. Challenges in using seaweed include high mineral content and difficulties in preservation and transportation. A potential preservation method is ensiling though current knowledge on its feasibility for animal feed is scarce. Future research has the potential to provide formulations that are palatable, bioactive and dose specific in providing optimised and sustainable livestock nutrition.

Society demands a more sustainable animal production system which, at the same time, needs to become more efficient to satisfy growing demand with increasingly limited resources. Within this context, we can identify two key trends in animal feed research:

- the development of real-time analytical techniques, and
- the search for novel ingredients.

In the short term, automated screening techniques ensuring feed safety, traceability and real-time evaluation of the ingredients entering the mill will become the rule in the feed industry. The combination of on-farm measurements (spectroscopy, (bio) sensors) with novel statistical analyses will help to fulfil these objectives by processing of a high quantity of data on a minute-by-minute basis.

Products initially considered unsuitable for efficient animal nutrition, as is the case with seaweed, insects or agri-food by-products, can now become common feed materials in the near future. They represent a valuable source of protein, polysaccharides, vitamins, minerals and other bioactive compounds. The application of methods to improve digestibility (enzymes, fermentation) and extraction of compounds of interest, combined with the precise assessment of their functionality by novel *in vitro* models, will boost their use as feed ingredients.

9 Where to look for further information

The following articles provide a good overview of the subject:

- Halmemies-Beauchet-Filleau, A., Rinne, M., Lamminen, M., Mapato, C., Ampapon, T., Wanapat, M. and Vanhatalo, A. (2018). Alternative and novel feeds for ruminants: nutritive value, product quality and environmental aspects. *Animal* 12(S2): s295–s309. DOI: 10.1017/S1751731118002252.

- Zhang, Q. W., Lin, L. G. and Ye, W. C. (2018). Techniques for extraction and isolation of natural products: A comprehensive review. *Chinese Medicine* 13(1). *BioMed Central*: 1-26. DOI: 10.1186/s13020-018-0177-x.
- Shahidi, F. and Zhong, Y. (2015). Measurement of antioxidant activity. *Journal of Functional Foods* 18: 757-781. DOI: 10.1016/j.jff.2015.01.047.
- Cowan, M. M. (1999). Plant products as antimicrobial agents. *Clinical Microbiology Reviews* 12(4): 564-582. DOI: 10.1128/cmr.12.4.564.

Key standard analytical methods in this area can be found at the following organisations:

- Association of Official Agricultural Chemists (AOAC) (www.aoac.org),
- American Oil Chemists' Society (AOCS) (www.aocs.org), and
- Clinical and Laboratory Standards Institute (IITA) (www.clsi.org/).

Some of the most relevant journals including novel analytical methods in the feed sector:

- Journal of Agricultural and Food Chemistry,
- Animal Feed Science and Technology,
- Industrial Crops and Products,
- Plant methods, and
- Precision Agriculture.

Key conferences:

- Recent Advantages in Food Analysis (RAFA) is one of the most prestigious international conferences on this field celebrated annually in Prague (Czech Republic). This symposium provides an updated overview of contemporary trends in analytical & bioanalytical strategies in food quality and safety control and discusses challenges and novel approaches in food and natural products analysis.
- European Federation of Animal Science annual meetings (EAAP) - this symposium represents the best opportunity to share the research at the most updated and scientifically advanced animal science forum in the world. The conference has a specific commission dedicated to Animal Nutrition in which many of the topics included in the chapter are discussed.

10 References

AOAC (Association of Official Analytical Chemists) (1990). *Official Methods of Analysis* (15th edn.). AOAC, Arlington, VA.

Alho, H. and Leinonen, J. (1999). Total antioxidant activity measured by chemiluminescence methods. *Methods in Enzymology* 299: 3–15. DOI: 10.1016/S0076-6879(99)99004-3.

Álvarez-Barrientos, A., Arroyo, J., Cantón, R., Nombela, C. and Sánchez-Pérez, M. (2000). Applications of flow cytometry to clinical microbiology. *Clinical Microbiology Reviews* 13(2): 167–195. DOI: 10.1128/CMR.13.2.167-195.2000.

Álvarez-Ordóñez, A., Carvajal, A., Arguello, H., Martínez-Lobo, F. J., Naharro, G. and Rubio, P. (2013). Antibacterial activity and mode of action of a commercial citrus fruit extract. *Journal of Applied Microbiology* 115(1): 50–60. DOI: 10.1111/jam.12216.

Ameer, K., Shahbaz, H. M. and Kwon, J. H. (2017). Green extraction methods for polyphenols from plant matrices and their byproducts: a review. *Comprehensive Reviews in Food Science and Food Safety* 16(2): 295–315. DOI: 10.1111/1541-4337.12253.

Apak, R., Güçlü, K., Özyürek, M. and Karademir, S. E. (2004). Novel total antioxidant capacity index for dietary polyphenols and vitamins C and E, using their cupric ion reducing capability in the presence of neocuproine: CUPRAC Method. *Journal of Agricultural and Food Chemistry* 52(26): 7970–7981.

Aquino, J. de S., Batista, K. S., Menezes, F. N. D. D., Lins, P., Gomes, J. and Silva, L. (2017). Models to evaluate the prebiotic potential of foods. In: *Functional Food–Improve Health through Adequate Food*. DOI: 10.5772/intechopen.69174.

Aufrère, J., Theodoridou, K., Mueller-Harvey, I., Yu, P. and Andueza, D. (2013). Ruminal dry matter and nitrogen degradation in relation to condensed tannin and protein molecular structures in sainfoin (Onobrychis viciifolia) and lucerne (Medicago sativa). *The Journal of Agricultural Science* 152: 333–345..

Balić, A., Vlašić, D., Žužul, K., Marinović, B. and Bukvić Mokos, Z. (2020). Omega-3 Versus Omega-6 polyunsaturated fatty acids in the prevention and treatment of inflammatory skin diseases. *International Journal of Molecular Sciences* 21(3): 741. DOI: 10.3390/ijms21030741.

Baltić, B., Starčević, M., Đorđević, J., Mrdović, B. and Marković, R. (2017). Importance of medium chain fatty acids in animal nutrition. *IOP Conference Series: Earth and Environmental Science* 85: 012048. DOI: 10.1088/1755-1315/85/1/012048.

Bindelle, J., Buldgen, A., Lambotte, D., et al. (2007). Effect of pig faecal donor and of pig diet composition on in vitro fermentation of sugar beet pulp. *Animal Feed Science and Technology*. DOI: 10.1016/j.anifeedsci.2006.03.010.

Birk, B., Staehle, A., Meier, M., Palm, M., Funk-Weyer, D., Breves, G. and Seulberger, H. (2018). Investigation of ruminant xenobiotic metabolism in a modified rumen simulation system (RUSITEC). *ALTEX* 35(3): 379–389. DOI: 10.14573/altex.1712221.

Bligh, E. G. and Dyer, W. J. (1959). A rapid method of total lipid extraction and purification. *Canadian Journal of Biochemistry and Physiology* 37(8): 911–917.

Boisen, S. and Fernández, J. A. (1997). Prediction of the total tract digestibility of energy in feedstuffs and pig diets by in vitro analyses. *Animal Feed Science and Technology* 68(3–4): 277–286. DOI: 10.1016/S0377-8401(97)00058-8.

Botsoglou, N. A., Fletouris, D. J., Papageorgiou, G. E., Vassilopoulos, V. N., Mantis, A. J. and Trakatellis, A. G. (1994). Rapid, sensitive and specific thiobarbituric acid method for measuring lipid peroxidation in animal tissue, food and feed samples. *Journal of Agricultural and Food Chemistry* 42: 1931–1937.

Brand-Williams, W., Cuvelier, M. E. and Berset, C. (1995). Use of a free radical method to evaluate antioxidant activity. *LWT–Food Science and Technology* 28(1): 25–30.

Bryan, D. D. S. L. and Classen, H. L. (2020). In vitro methods of assessing protein quality for poultry. *Animals (Basel)* 10(4): 551. DOI: 10.3390/ani10040551.

Butler, G. (2014). Manipulating dietary PUFA in animal feed: implications for human health. *Proceedings of the Nutrition Society* 73(1): 87–95. DOI: 10.1017/S0029665113003790.

Campbell, M., Ortuño, J., Ford, L., Davies, D. R., Koidis, A., Walsh, P. J. and Theodoridou, K. (2020). The effect of ensiling on the nutritional composition and fermentation characteristics of brown seaweeds as a ruminant feed ingredient. *Animals: An Open Access Journal from MDPI* 10(6): 1–17. DOI: 10.3390/ani10061019.

Charoensiddhi, S., Conlon, M. A., Vuaran, M. S., Franco, C. M. M. and Zhang, W. (2016). Impact of extraction processes on prebiotic potential of the brown seaweed Ecklonia radiata by in vitro human gut bacteria fermentation. *Journal of Functional Foods* 24: 221–230. DOI: 10.1016/j.jff.2016.04.016.

Chen, G., Chen, K., Zhang, R., Chen, X., Hu, P. and Kan, J. (2018a). Polysaccharides from bamboo shoots processing by-products: new insight into extraction and characterization. *Food Chemistry* 245: 1113–1123. DOI: 10.1016/j.foodchem.2017.11.059.

Chen, H., Wierenga, P., Hendriks, W. and Jansman, A. (2018b). In vitro protein digestion kinetics of protein sources for pigs. *Animal* 13: 1–11. DOI: 10.1017/S1751731118002811.

Chilliard, Y., Bauchart, D., Lessire, M., Schmidely, P. and Mourot, J. (2008). Qualité des produits: modulation par l'alimentation des animaux de la composition en acides gras du lait et de la viande. *INRA Productions Animales* 21(1): 95–106.

Chiu, H.-H. and Kuo, C. H. (2020). Gas chromatography-mass spectrometry-based analytical strategies for fatty acid analysis in biological samples. *Journal of Food and Drug Analysis* 28(1): 60–73. DOI: 10.1016/j.jfda.2019.10.003.

CLSI (2015). *Performance Standards for Antimicrobial Disk and Dilution Susceptibility Tests for Bacteria Isolated from Animals* (3rd edn.). Clinical and Laboratory Standards Institute, Wayne, PA.

Connan, S., Goulard, F., Stiger, V., Deslandes, E. and Gall, E. A. (2004). Interspecific and temporal variation in phlorotannin levels in an assemblage of brown algae. *Botanica Marina* 47: 410–416.

Cowan, M. M. (1999). Plant products as antimicrobial agents. *Clinical Microbiology Reviews* 12(4): 564–582. DOI: 10.1128/cmr.12.4.564.

Craft, B. D., Kerrihard, A. L., Amarowicz, R. and Pegg, R. B. (2012). Phenol-based antioxidants and the in vitro methods used for their assessment. *Comprehensive Reviews in Food Science and Food Safety* 11(2): 148–173. DOI: 10.1111/j.1541-4337.2011.00173.x.

Deng, Y. T., Liang, G., Shi, Y., Li, H., Zhang, J., Mao, X., Fu, Q., Peng, W., Chen, Q. and Shen, D. (2016). Condensed tannins from Ficus altissima leaves: structural, antioxidant, and antityrosinase properties. *Process Biochemistry* 51(8): 1092–1099. DOI: 10.1016/j.procbio.2016.04.022.

Deyrieux, C., Villeneuve, P., Baréa, B., Decker, E. A., Guiller, I., Michel Salaun, F. and Durand, E. (2018). Measurement of peroxide values in oils by triphenylphosphine/triphenylphosphine oxide (TPP/TPPO) assay coupled with FTIR-ATR spectroscopy: comparison with iodometric titration. *European Journal of Lipid Science and Technology* 120(8). DOI: 10.1002/ejlt.201800109: 1800109.

Dinis, T. C. P., Madeira, V. M. C. and Almeida, L. M. (1994). Action of phenolic derivatives (acetaminophen, salicylate, and 5-aminosalicylate) as inhibitors of membrane lipid peroxidation and as peroxyl radical scavengers. *Archives of Biochemistry and Biophysics* 315(1): 161–169. DOI: 10.1006/abbi.1994.1485.

Duarte, A. C., Holman, D. B., Alexander, T. W., Durmic, Z., Vercoe, P. E. and Chaves, A. V. (2017a). The type of forage substrate preparation included as substrate in a RUSITEC system affects the ruminal microbiota and fermentation characteristics . *Frontiers in Microbiology* 8: 704. DOI: 10.3389/fmicb.2017.00704.

Duarte, F. N. D., Rodrigues, J. B., da Costa Lima, M., Lima, M. D. S., Pacheco, M. T. B., Pintado, M. M. E., de Souza Aquino, J. and de Souza, E. L. (2017b). Potential prebiotic properties of cashew apple (Anacardium occidentale L.) agro-industrial byproduct on Lactobacillus species. *Journal of the Science of Food and Agriculture* 97(11): 3712–3719. DOI: 10.1002/jsfa.8232.

Elleuch, L., Shaaban, M., Smaoui, S., Mellouli, L., Karray-Rebai, I., Fourati-Ben Fguira, L., Shaaban, K. A. and Laatsch, H. (2010). Bioactive secondary metabolites from a new terrestrial Streptomyces sp. TN262. *Applied Biochemistry and Biotechnology* 162(2): 579–593. DOI: 10.1007/s12010-009-8808-4.

Evans, F. D. and Critchley, A. T. (2014). Seaweeds for animal production use. *Journal of Applied Phycology* 26(2): 891–899. DOI: 10.1007/s10811-013-0162-9.

Fahey, G. C. Jr., Novotny, L., Layton, B. and Mertens, D. R. (2018 Aug 2). Critical factors in determining fiber content of feeds and foods and their ingredients. *Journal of AOAC International*. DOI: 10.5740/jaoacint.18-0067. Epub ahead of print.

Farmer, E. H. and Sutton, D. A. (1943). The course of autoxidation reactions in polyisoprenes and allied compounds. Part IV: the isolation and constitution of photochemically-formed methyl oleate peroxide. *Journal of the Chemical Society*. DOI: 10.1039/jr9430000119.

Fellenberg, M. A. and Speisky, H. (2006). Antioxidants: their effects on broiler oxidative stress and its meat oxidative stability. *World's Poultry Science Journal* 62(1): 53–70. DOI: 10.1079/WPS200584.

Feng, H. L., Tian, L., Chai, W. M., Chen, X. X., Shi, Y., Gao, Y. S., Yan, C. L. and Chen, Q. X. (2014). Isolation and purification of condensed tannins from flamboyant tree and their antioxidant and antityrosinase activities. *Applied Biochemistry and Biotechnology* 173(1): 179–192. DOI: 10.1007/s12010-014-0828-z.

Fennema, O. R. (1996). *Food Chemistry*, 3rd edn. Marcel Dekker, New York, NY.

Ferreyra, S. G., Antoniolli, A., Bottini, R. and Fontana, A. (2020). Bioactive compounds and total antioxidant capacity of cane residues from different grape varieties. *Journal of the Science of Food and Agriculture* 100(1): 376–383. DOI: 10.1002/jsfa.10065.

Folch, J., Lees, M. and Stanley, G. H. S. (1957). A simple method for the isolation and purification of total lipides from animal tissues. *The Journal of Biological Chemistry* 226(1): 497–509.

Ford, L., Stratakos, A. C., Theodoridou, K., Dick, J. T. A., Sheldrake, G. N., Linton, M., Corcionivoschi, N. and Walsh, P. J. (2020). Polyphenols from brown seaweeds as a potential antimicrobial agent in animal feeds. *ACS Omega* 5(16): 9093–9103. DOI: 10.1021/acsomega.9b03687.

Ford, L., Theodoridou, K., Sheldrake, G. N. and Walsh, P. J. (2019). A critical review of analytical methods used for the chemical characterisation and quantification of phlorotannin compounds in brown seaweeds. *Phytochemical Analysis: PCA* 30(6): 587–599.

Fox, D. G., Tedeschi, L. O., Tylutki, T. P., Russell, J. B., Van Amburgh, M. E., Chase, L. E., Pell, A. N. and Overton, T. R. (2004). The Cornell Net carbohydrate and Protein System model for evaluating herd nutrition and nutrient excretion. *Animal Feed Science and Technology* 112(1–4): 29–78.

Fuentes-Alventosa, J. M., Jaramillo-Carmona, S., Rodríguez-Gutiérrez, G., Guillén-Bejarano, R., Jiménez-Araujo, A., Fernández-Bolaños, J. and Rodríguez-Arcos, R. (2013). Preparation of bioactive extracts from asparagus by-product. *Food and Bioproducts Processing* 91(2): 74–82. DOI: 10.1016/j.fbp.2012.12.004.

Gea, A., Stringano, E., Brown, R. H. and Mueller-Harvey, I. (2011). In situ analysis and structural elucidation of sainfoin (Onobrychis viciifolia) tannins for high-throughput germplasm screening. *Journal of Agricultural and Food Chemistry* 59(2): 495–503. DOI: 10.1021/jf103609p.

Getachew, G., DePeters, E. J. and Robinson, P. H. (2004). In vitro gas production provides effective method for assessing ruminant feeds. *California Agriculture* 58: 54–58.

Gidenne, T. (2014). Dietary fibres: their analysis in animal feeding, and their role in rabbit nutrition and health. Indonesian. *Bulletin of Animal and Veterinary Sciences* 23(10): 14334/wartazoa.v23i4.1010.

Gómez, B., Gullón, B., Remoroza, C., Schols, H. A., Parajó, J. C. and Alonso, J. L. (2014). Purification, characterization, and prebiotic properties of pectic oligosaccharides from orange peel wastes. *Journal of Agricultural and Food Chemistry* 62(40): 9769–9782. DOI: 10.1021/jf503475b.

Gonçalves, S. and Romano, A. (2017). Inhibitory properties of phenolic compounds Against enzymes linked with human diseases. In: *Phenolic Compounds - Biological Activity*. DOI: 10.5772/66844.

Grabber, J. H. and Zeller, W. E. (2020). Direct versus sequential analysis of procyanidin- a nd prodelphinidin-based condensed tannins by the HCl-butanol-acetone-iron assay. *Journal of Agricultural and Food Chemistry* 68(10): 2906–2916. DOI: 10.1021/acs.jafc.9b01307.

Guil-Guerrero, J. L., Ramos, L., Moreno, C., Zúñiga-Paredes, J. C., Carlosama-Yépez, M. and Ruales, P. (2016). Plant-food by-products to improve farm-animal health. *Animal Feed Science and Technology* 220: 121–135. DOI: 10.1016/j.anifeedsci.2016.07.016.

Gullón, B., Pintado, M. E., Fernández-López, J., Pérez-Álvarez, J. A. and Viuda-Martos, M. (2015). In vitro gastrointestinal digestion of pomegranate peel (Punica granatum) flour obtained from co-products: changes in the antioxidant potential and bioactive compounds stability. *Journal of Functional Foods* 19: 617–628. DOI: 10.1016/j.jff.2015.09.056.

Habeebullah, K. H. and Jacobsen, C. (2015). Antioxidant activity of seaweed extracts: in vitro assays, evaluation in 5% fish oil-in-water emulsions and characterization. *Journal of the American Oil Chemists' Society* 92(4): 571–587. DOI: 10.1007/s11746-015-2624-5.

Habeebullah, S. F., Alagarsamy, S., Sattari, Z., Al-Haddad, S., Fakhraldeen, S., Al-Ghunaim, A. and Al-Yamani, F. (2020). Enzyme-assisted extraction of bioactive compounds from brown seaweeds and characterization. *Journal of Applied Phycology* 32(1): 615–629. DOI: 10.1007/s10811-019-01906-6.

Hamed, M., Bougatef, H., Karoud, W., Krichen, F., Haddar, A., Bougatef, A. and Sila, A. (2020). Polysaccharides extracted from pistachio external hull: characterization, antioxidant activity and potential application on meat as preservative. *Industrial Crops and Products* 148. DOI: 10.1016/j.indcrop.2020.112315: 112315.

Holden, L. A. (1999). Comparison of methods of in vitro dry matter digestibility for ten feeds. *Journal of Dairy Science* 82(8): 1791–1794.

Jha, R., Bindelle, J., Van Kessel, A. and Leterme, P. (2011). In vitro fibre fermentation of feed ingredients with varying fermentable carbohydrate and protein levels and protein synthesis by colonic bacteria isolated from pigs. *Animal Feed Science and Technology* 165(3–4): 191–200. DOI: 10.1016/j.anifeedsci.2010.10.002.

Jha, R., Fouhse, J. M., Tiwari, U. P., Li, L. and Willing, B. P. (2019). Dietary fiber and intestinal health of monogastric animals. *Frontiers in Veterinary Science* 6: 48. DOI: 10.3389/fvets.2019.00048.

Jha, R., Owusu-Asiedu, A., Simmins, P. H., Pharazyn, A. and Zijlstra, R. T. (2012). Degradation and fermentation characteristics of wheat coproducts from flour milling in the pig intestine studied in vitro. *Journal of Animal Science* 90(Suppl. 4): 173–175. DOI: 10.2527/jas.53874.

Jordán, M. J., Lax, V., Rota, M. C., Lorán, S. and Sotomayor, J. A. (2013). Effect of the phenological stage on the chemical composition, and antimicrobial and antioxidant properties of Rosmarinus officinalis L essential oil and its polyphenolic extract. *Industrial Crops and Products* 48: 144–152. DOI: 10.1016/j.indcrop.2013.04.031.

Julkunen-tiitto, R. (1985). Phenolic constituents in the leaves of northern willows: methods for the analysis of certain phenolics. *Journal of Agricultural and Food Chemistry* 33(2): 213–217. DOI: 10.1021/jf00062a013.

Kampa, M., Nistikaki, A., Tsaousis, V., Maliaraki, N., Notas, G. and Castanas, E. (2002). A new automated method for the determination of the Total Antioxidant Capacity (TAC) of human plasma, based on the crocin bleaching assay. *BMC Clinical Pathology* 2(1): 3. DOI: 10.1186/1472-6890-2-3.

Kennedy, J. A. and Jones, G. P. (2001). Analysis of proanthocyanidin cleavage products following acid-catalysis in the presence of excess phloroglucinol. *Journal of Agricultural and Food Chemistry* 49(4): 1740–1746. DOI: 10.1021/jf001030o.

Kerr, B. J. and Shurson, G. C. (2013). Strategies to improve fiber utilization in swine. *Journal of Animal Science and Biotechnology* 4(1): 11.

Kinley, R. D. and Fredeen, A. H. (2016). In vitro evaluation of feeding North Atlantic stormtoss seaweeds on ruminal digestion. *Journal of Applied Phycology* 27(6): 2387–2393. DOI: 10.1007/s10811-014-0487-z.

Kumar, C. S., Ganesan, P., Suresh, P. V. and Bhaskar, N. (2008). Seaweeds as a source of nutritionally beneficial compounds–a review. *Journal of Food Science and Technology* 45: 1–13.

Laguerre, M., Lecomte, J. and Villeneuve, P. (2007). Evaluation of the ability of antioxidants to counteract lipid oxidation: existing methods, new trends and challenges. *Progress in Lipid Research* 46(5): 244–282. DOI: 10.1016/j.plipres.2007.05.002.

Lee, F. Y., Htar, T. T. and Akowuah, G. A. (2015). ATR-FTIR and spectrometric methods for the assay of crocin in commercial saffron spices (Crocus Savitus L.). *International Journal of Food Properties* 18(8): 1773–1783. DOI: 10.1080/10942912.2014.923911.

Leiva, A. and Granados-Chinchilla, F. (2020). Fatty acid profiling in animal feeds and related food matrixes using a fast GC/MS method and in situ derivatization. *International Journal of Agriculture, Environment and Food Sciences* 4(1): 70–89.

Lichtenthaler, H. K. (1987). [34] Chlorophylls and carotenoids: pigments of photosynthetic biomembranes. *Methods in Enzymology*: 350–382. DOI: 10.1016/0076-6879(87)48036-1.

Liu, M., Zhang, L., Ser, S. L., Cumming, J. R. and Ku, K. M. (2018). Comparative phytonutrient analysis of broccoli by-products: the potentials for broccoli by-product utilization. *Molecules* 23(4). DOI: 10.3390/molecules23040900.

Liyanaarachchi, G. V. V., Mahanama, K. R. R., Somasiri, H. P. P. S., Punyasiri, P. A. N. and Kottawa-Arachchi, J. D. (2020). Analysis of total amino acids in rice using a validated reversed-phase high performance liquid chromatographic method with diode array detection (RP-HPLCDAD). *Int. J. Analyt. Bioanalyt Methods* 2: 008.

López-Bascón, M. A. and Luque de Castro, M. D. (2019). Soxhlet extraction. In: *Liquid-Phase Extraction*. DOI: 10.1016/B978-0-12-816911-7.00011-6.

Lussignoli, S., Fraccaroli, M., Andrioli, G., Brocco, G. and Bellavite, P. (1999). A microplate-based colorimetric assay of the total peroxyl radical trapping capability of human plasma. *Analytical Biochemistry* 269(1): 38–44. DOI: 10.1006/abio.1999.4010.

Mæhre, H. K., Dalheim, L., Edvinsen, G. K., Elvevoll, E. O. and Jensen, I.-J. (2018). Protein determination-method matters. *Foods* 7(1): 5. DOI: 10.3390/foods7010005.

Maia, M. R., Fonseca, A. J., Oliveira, H. M., Mendonça, C. and Cabrita, A. R. (2016) The potential role of seaweeds in the natural manipulation of rumen fermentation and methane production. *Scientific Reports* 6: 32321.

Makkar, H. P. S. (2000). *Quantification of Tannins in Tree Foliage. A Laboratory Manual for the FAO/IAEA*. Academic Publ Dordrecht, The Netherlands

Mariotti, F., Tomé, D. and Mirand, P. P. (2008). Converting nitrogen into protein–Beyond 6.25 and Jones' factors. *Critical Reviews in Food Science and Nutrition* 48(2): 177-184. DOI: 10.1080/10408390701279749.

Mejri, F., Baati, T., Martins, A., Selmi, S., Luisa Serralheiro, M., Falé, P. L., Rauter, A., Casabianca, H. and Hosni, K. (2020). Phytochemical analysis and in vitro and in vivo evaluation of biological activities of artichoke (Cynara Scolymus L.) floral stems: towards the valorization of food by-products. *Food Chemistry* 333: 127506. DOI: 10.1016/j.foodchem.2020.127506.

Mgbeahuruike, E. E., Fyhrquist, P., Vuorela, H., Julkunen-Tiitto, R. and Holm, Y. (2018). Alkaloid-rich crude extracts, fractions and piperamide alkaloids of piper guineense possess promising antibacterial effects. *Antibiotics* 7(4): 1–18. DOI: 10.3390/antibiotics7040098.

Moudache, M., Silva, F., Nerín, C. and Zaidi, F. (2020). Olive cake and leaf extracts as valuable sources of antioxidant and antimicrobial compounds: a comparative study. *Waste and Biomass Valorization* 12(3): 1431–1445. DOI: 10.1007/s12649-020-01080-8.

Murphy, D., Ricci, A., Auce, Z., Beechinor, J. G., Bergendahl, H., Breathnach, R., Bureš, J., Duarte Da Silva, J. P., Hederová, J., Hekman, P., Ibrahim, C., Kozhuharov, E., Kulcsár, G., Lander Persson, E., Lenhardsson, J. M., Mačiulskis, P., Malemis, I., Markus-Cizelj, L., Michaelidou-Patsia, A., Nevalainen, M., Pasquali, P., Rouby, J. C., Schefferlie, J., Schlumbohm, W., Schmit, M., Spiteri, S., Srčič, S., Taban, L., Tiirats, T., Urbain, B., Vestergaard, E. M., Wachnik-Święcicka, A., Weeks, J., Zemann, B., Allende, A., Bolton, D., Chemaly, M., Fernandez Escamez, P. S., Girones, R., Herman, L., Koutsoumanis, K., Lindqvist, R., Nørrung, B., Robertson, L., Ru, G., Sanaa, M., Simmons, M., Skandamis, P., Snary, E., Speybroeck, N., Ter Kuile, B., Wahlström, H., Baptiste, K., Catry, B., Cocconcelli, P. S., Davies, R., Ducrot, C., Friis, C., Jungersen, G., More, S., Muñoz Madero, C., Sanders, P., Bos, M., Kunsagi, Z., Torren Edo, J., Brozzi, R., Candiani, D., Guerra, B., Liebana, E., Stella, P., Threlfall, J. and Jukes, H.; EMA Committee for Medicinal Products for Veterinary Use (CVMP) and EFSA Panel on Biological Hazards (BIOHAZ) (2017). EMA and EFSA Joint Scientific Opinion on measures to reduce the need to use antimicrobial agents in animal husbandry in the European Union, and the resulting impacts on food safety (RONAFA). *EFSA Journal. European Food Safety Authority* 15(1): e04666. DOI: 10.2903/j.efsa.2017.4666.

NRC (2007). *Nutrient Requirements of Horses* (6th rev. edn.). Natl. Academy Press, Washington, DC.

Ortuño, J., Serrano, R. and Bañón, S. (2015). Antioxidant and antimicrobial effects of dietary supplementation with rosemary diterpenes (carnosic acid and carnosol) vs vitamin E on lamb meat packed under protective atmosphere. *Meat Science* 110: 62-69. DOI: 10.1016/j.meatsci.2015.07.011.

Ortuño, J., Serrano, R. and Bañón, S. (2016). Use of dietary rosemary diterpenes to inhibit rancid volatiles in lamb meat packed under protective atmosphere. *Animal: An International Journal of Animal Bioscience* 10(8): 1391-1401. DOI: 10.1017/S1751731116000392.

Owen, E., Jayasuriya, M. C. N., Hamilton, R. and Lalenta, M. (1991). Use of a long-term rumen simulation technique (Rusitec) to provide micro-organisms for in vitro digestibility assays. *The Journal of Agricultural Science* 116(2): 297-301. DOI: 10.1017/S0021859600077716.

Panzella, L., Moccia, F., Nasti, R., Marzorati, S., Verotta, L. and Napolitano, A. (2020). Bioactive phenolic compounds from agri-food wastes: an update on green and sustainable extraction methodologies. *Frontiers in Nutrition* 7(May): 60. DOI: 10.3389/fnut.2020.00060.

Pfaller, M. A., Sheehan, D. J. and Rex, J. H. (2004). Determination of fungicidal activities against yeasts and molds: lessons learned from bactericidal testing and the need for standardization. *Clinical Microbiology Reviews* 17(2): 268-280. DOI: 10.1128/CMR.17.2.268-280.2004.

Prior, R. L., Hoang, H., Gu, L., Wu, X., Bacchiocca, M., Howard, L., Hampsch-Woodill, M., Huang, D., Ou, B. and Jacob, R. (2003). Assays for hydrophilic and lipophilic antioxidant capacity (oxygen radical absorbance capacity (ORACFL)) of plasma and other biological and food samples. *Journal of Agricultural and Food Chemistry* 51(11): 3273-3279. DOI: 10.1021/jf0262256.

Prommaban, A., Utama-Ang, N., Chaikitwattana, A., Uthaipibull, C., Porter, J. B. and Srichairatanakool, S. (2020). Phytosterol, lipid and phenolic composition, and biological activities of guava seed oil. *Molecules* 25(11). DOI: 10.3390/molecules25112474.

Pulido, R., Bravo, L. and Saura-Calixto, F. (2000). Antioxidant activity of dietary polyphenols as determined by a modified ferric reducing/antioxidant power assay. *Journal of Agricultural and Food Chemistry* 48(8): 3396-3402. DOI: 10.1021/jf9913458.

Re, R., Pellegrini, N., Proteggente, A., Pannala, A., Yang, M. and Rice-Evans, C. (1999). Antioxidant activity applying an improved ABTS radical cation decolorization assay. *Free Radical Biology and Medicine* 26(9-10): 1231-1237.

Regulation (EC) No 1831/2003 of the European Parliament and the Council of the European Union of 22 September 2003 on Additives for Use in Animal Nutrition.

Rohn, S., Rawel, H. M. and Kroll, J. (2002). Inhibitory effects of plant phenols on the activity of selected enzymes. *Journal of Agricultural and Food Chemistry* 50(12): 3566-3571. DOI: 10.1021/jf011714b.

Rusu, M. E., Fizeșan, I., Pop, A., Gheldiu, A. M., Mocan, A., Crișan, G., Vlase, L., Loghin, F., Popa, D. S. and Tomuta, I. (2019). Enhanced recovery of antioxidant compounds from hazelnut (Corylus avellana l.) involucre based on extraction optimization: phytochemical profile and biological activities. *Antioxidants* 8(10). DOI: 10.3390/antiox8100460.

Sahariah, P. and Másson, M. (2017). Antimicrobial chitosan and chitosan derivatives: a review of the structure-activity relationship. *Biomacromolecules* 18(11): 3846-3868. DOI: 10.1021/acs.biomac.7b01058.

Shahidi, F. and Zhong, Y. (2015). Measurement of antioxidant activity. *Journal of Functional Foods* 18: 757–781. DOI: 10.1016/j.jff.2015.01.047.

Shantha, N. C. and Decker, E. A. (1994). Rapid, sensitive, iron-based spectrophotometric methods for determination of peroxide values of food lipids. *Journal of AOAC International* 77(2): 421–424.

Shimada, K., Fujikawa, K., Yahara, K. and Nakamura, T. (1992). Antioxidative properties of xanthan on the autoxidation of soybean oil in cyclodextrin emulsion. *Journal of Agricultural and Food Chemistry* 40(6): 945–948. DOI: 10.1021/jf00018a005.

Silanikove, N., Shinder, D., Gilboa, N., et al. (1996). Binding of poly(ethylene glycol) to samples of forage plants as an assay of tannins and their negative effects on ruminal degradation. *Journal of Agricultural and Food Chemistry*. DOI: 10.1021/jf9602277.

Simon, O. and Igbasan, F. (2002). In vitro properties of phytases from various microbial origins. *International Journal of Food Science and Technology* 37(7): 813–822. DOI: 10.1046/j.1365-2621.2002.00621.x.

Singleton, V. L. and Rossi, J. A. J. (1965). Colorimetry to total phenolics with phosphomolybdic acid reagents. *American Journal of Enology and Viniculture* 16: 144–158.

Smith, P. E., Waters, S. M., Kenny, D. A., Boland, T. M., Heffernan, J. and Kelly, A. K. (2020). Replacing barley and soybean meal with by-products, in a pasture based diet, alters daily methane output and the rumen microbial community *in vitro* using the Rumen Simulation Technique (RUSITEC). *Frontiers in Microbiology* 11(11): 1614. DOI: 10.3389/fmicb.2020.01614.

Stratakos, A. C. and Koidis, A. (2016). Methods for extracting essential oils. In *Essential Oils in Food Preservation, Flavor and Safety*. Elsevier Inc. DOI: 10.1016/B978-0-12-416641-7.00004-3.

Szabo, K., Dulf, F. V., Diaconeasa, Z. and Vodnar, D. C. (2019). Antimicrobial and antioxidant properties of tomato processing byproducts and their correlation with the biochemical composition. *LWT* 116. DOI: 10.1016/j.lwt.2019.108558.

Szkudzińska, K., Smutniak, I., Rubaj, J., Korol, W. and Bielecka, G. (2017). Method validation for determination of amino acids in feed by UPLC. *Accreditation and Quality Assurance* 22(5): 247–252. DOI: 10.1007/s00769-017-1281-9.

Theodoridou, K., Aufrère, J., Andueza, D., Pourrat, J., Le Morvan, A., Stringano, E., Mueller-Harvey, I. and Baumont, R. (2010). Effects of condensed tannins in fresh sainfoin (Onobrychis viciifolia) on in vivo and in situ digestion in sheep. *Animal Feed Science and Technology* 160(1–2): 23–38. DOI: 10.1016/j.anifeedsci.2010.06.007.

Theodoridou, K. and Yu, P. (2013). Metabolic characteristics of the proteins in yellow-seeded and brown-seeded canola meal and presscake in dairy cattle: comparison of three systems (PDI, DVE, and NRC) in nutrient supply and feed milk value (FMV). *Journal of Agricultural and Food Chemistry* 61(11): 2820–2830. DOI: 10.1021/jf305171z.

Theodoridou, M. K., Williams, B. A., Dhanoa, M. S., McAllan, A. B. and France, J. (1994). A simple gas production method using a pressure transducer to determine the fermentation kinetics of ruminant feeds. *Animal Feed Science and Technology* 48(3-4): 185–197. DOI: 10.1016/0377-8401(94)90171-6.

Tremocoldi, M. A., Rosalen, P. L., Franchin, M., Massarioli, A. P., Denny, C., Daiuto, ÉR., Paschoal, J. A. R., Melo, P. S. and Alencar, S. M. (2018). Exploration of avocado by-products as natural sources of bioactive compounds. *PLoS ONE* 13(2): e0192577. DOI: 10.1371/journal.pone.0192577.

Uyeno, Y., Shigemori, S. and Shimosato, T. (2015). Effect of probiotics/prebiotics on cattle health and productivity. *Microbes and Environments* 30(2): 126–132. DOI: 10.1264/jsme2.ME14176.

Valgas, C., De Souza, S. Md, Smânia, E. F. A. and Smânia, A. (2007). Screening methods to determine antibacterial activity of natural products. *Brazilian Journal of Microbiology* 38(2): 369–380. DOI: 10.1590/S1517-83822007000200034.

Vargas, J. E., Andrés, S., López-Ferreras, L., Snelling, T. J., Yáñez-Ruíz, D. R., García-Estrada, C. and López, S. (2020). Dietary supplemental plant oils reduce methanogenesis from anaerobic microbial fermentation in the rumen. *Scientific Reports* 10(1): 1613. DOI: 10.1038/s41598-020-58401-z.

Van Soest, P. J. (1963). The use of detergents in the analysis of fibrous feeds: II. A rapid method for the determination of fiber and lignin. *Official Agriculture Chemistry* 46: 829.

Vella, F. M., De Masi, L., Calandrelli, R., Morana, A. and Laratta, B. (2019). Valorization of the agro-forestry wastes from Italian chestnut cultivars for the recovery of bioactive compounds. *European Food Research and Technology* 245(12): 2679–2686. DOI: 10.1007/s00217-019-03379-w.

Wang, Y. and McAllister, A. (2011). Brown algae as a feed additive: nutritional and health impacts on ruminant–a review. *Animal Feed: Types, Nutrition and Safety*: 1–32.

Wiegand, I., Hilpert, K. and Hancock, R. E. W. (2008). Agar and broth dilution methods to determine the minimal inhibitory concentration (MIC) of antimicrobial substances. *Nature Protocols* 3(2): 163–175. DOI: 10.1038/nprot.2007.521.

World Health Organization (2014). *The Evolving Threat of Antimicrobial Resistance: Options for Action*. WHO Publications, France.

Yahia, Y., Benabderrahim, M. A., Tlili, N., Bagues, M. and Nagaz, K. (2020). Bioactive compounds, antioxidant and antimicrobial activities of extracts from different plant parts of two Ziziphus Mill. species. *PLoS ONE* 15(5): e0232599. DOI: 10.1371/journal.pone.0232599.

Yu, P. (2004). Application of advanced synchrotron-based Fourier transform infrared microspectroscopy (SR-FTIR) to animal nutrition and feed science: a novel approach. *British Journal of Nutrition* 92(6): 869–885.

Yu, P. (2005). Protein secondary structures (α-helix and β-sheet) at a cellular level and protein fractions in relation to rumen degradation behaviours of protein: a new approach. *British Journal of Nutrition* 94(5): 655–665.

Yu, P. and Nuez-Ortín, W. G. (2010). Relationship of protein molecular structure to metabolisable proteins in different types of dried distillers grains with solubles: a novel approach. *British Journal of Nutrition* 104(10): 1429–1437.

Zeller, W. E. (2019). Activity, purification, and analysis of condensed tannins: current state of affairs and future endeavors. *Crop Science* 59(3): 886–904. DOI: 10.2135/cropsci2018.05.0323.

Zeller, W. E., Sullivan, M. L., Mueller-Harvey, I., Grabber, J. H., Ramsay, A., Drake, C. and Brown, R. H. (2015). Protein precipitation behavior of condensed tannins from Lotus pedunculatus and Trifolium repens with different mean degrees of polymerization. *Journal of Agricultural and Food Chemistry* 63(4): 1160–1168. DOI: 10.1021/jf504715p.

Zhang, G., Hu, M., He, L., Fu, P., Wang, L. and Zhou, J. (2013). Optimization of microwave-assisted enzymatic extraction of polyphenols from waste peanut shells and evaluation of its antioxidant and antibacterial activities in vitro. *Food and Bioproducts Processing* 91(2): 158–168. DOI: 10.1016/j.fbp.2012.09.003.

Zhang, Q. W., Lin, L. G. and Ye, W. C. (2018). Techniques for extraction and isolation of natural products: a comprehensive review. *Chinese Medicine* 13(1): 20. DOI: 10.1186/s13020-018-0177-x.

Zhishen, J., Mengcheng, T. and Jianming, W. (1999). The determination of flavonoid contents in mulberry and their scavenging effects on superoxide radicals. *Food Chemistry* 64(4): 555–559. DOI: 10.1016/S0308-8146(98)00102-2.

Chapter 2

Effect of processing techniques on the quality of animal feed

Dennis Forte, Dennis Forte & Associates Pty Ltd, Australia

1 Introduction

Nutritionists often focus upon the nutritional quality of raw materials (as measured via in vitro digestibility) and then assemble these raw materials within a formulation (or diet) so that they meet specific nutritional requirements for a particular animal species. When feeding mash diets or steam pelleted feed, it is assumed that there is minimal change in the supply of nutrients from the component ingredients through to the feed. However, when ingredients are processed via extrusion, the chemical and physical interactions that occur between ingredients during the extrusion cooking process alter the properties of individual components.

Nutritionists working with feed produced from extrusion cooking need to be aware of the influence of process parameters on the nutritional quality of the feed, including the feed conversion ratio (FCR). The primary objective of this chapter is to demonstrate how the manipulation of extrusion processing parameters such as specific mechanical energy (SME) can directly influence the nutritional value of a given formulation component such as starch in a dry animal feed formulation.

2 Extrusion cooking: an overview

An overview of a typical extrusion cooking process is shown in Fig. 1 (1). Key components of the extrusion process typically include:

http://dx.doi.org/10.19103/AS.2020.0083.08

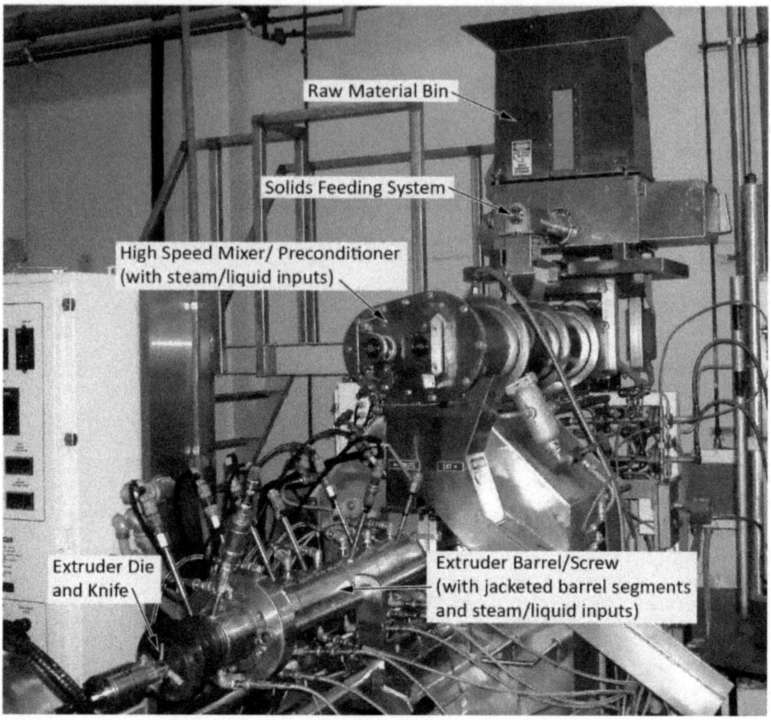

Figure 1 The key components of a typical extrusion cooking process.

- the feed bin;
- the feeder (which controls the processing rate);
- the preconditioner (where steam is added);
- the extruder (which imparts the shear); and
- the die and the cutter.

Down-stream ancillary equipment items such as dryers and coaters are also key components of the process but are not discussed here.

In simple terms, the transformation (or cooking) of a set of raw materials into a particular product involves the incorporation of a carefully defined and controlled amount of energy. The means by which the energy is imparted is defined by the rheological properties of the formulation, the design of the processing equipment, the operational parameters of the process and the desired physical attributes (e.g. shape, size and density) and chemical attributes (flavour, solubility, etc.) of the finished product (2).

There are three mechanisms for energy transfer in the cooking process:

- mechanical energy (also referred to as viscous dissipation);

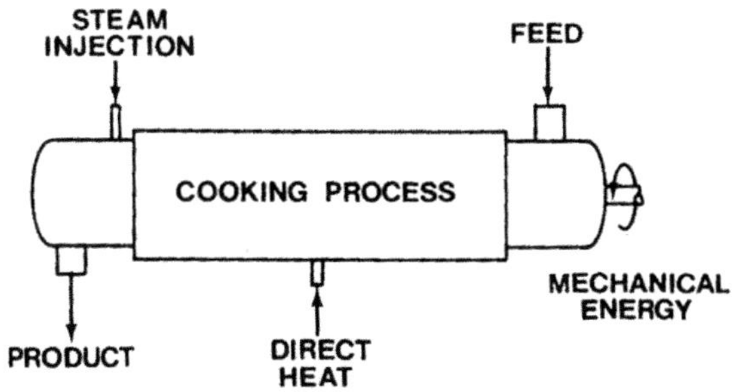

Figure 2 The generic cooking process (showing the main energy inputs).

- convective energy (associated with latent heat transfer); and
- conductive energy (associated with indirect heat transfer).

This process is shown in Fig. 2. The manipulation of these energy inputs has a direct influence upon product attributes. The term gelatinization is often used to describe the cooking process for starch-based formulations. True 'gelatinization' is only possible in the presence of excess water (i.e. at a total moisture content, $w > 60\%$ w/w). This is, however, only one of a number of alternative mechanisms (i.e. at $w < 60\%$ w/w) that achieve cooking. Product attributes such as flavour, colour and texture are the direct result of the pathway chosen.

3 Processing of cereals for food and feed

Miller (3) discusses the broad range of cereal cooking (or processing) conditions that are currently used in the food industry. The key variables are time, temperature and moisture content during processing. These conditions are shown in Fig. 3. This shows that the role of (in process) moisture content is critical. The use of short-time high-temperature processing conditions (typically with a high shear rate) produces 'low' moisture conditions and products, whilst the opposite produces 'high' moisture conditions and products. Physical attributes of the finished product (e.g. pellet density, hardness and durability are also directly influenced by these processing parameters.

During cooking there are numerous potential pathways for starch degradation (4). These are shown in Fig. 4. The various pathways may be defined in terms of the level of mechanical energy input into the process. The two extremes may be defined as:

- Mechanical degradation where both the thermal energy input (both convective and conductive) and the water content are minimized. A typical example of this type of process is the preparation of direct expanded snack foods using high shear collet extruders.

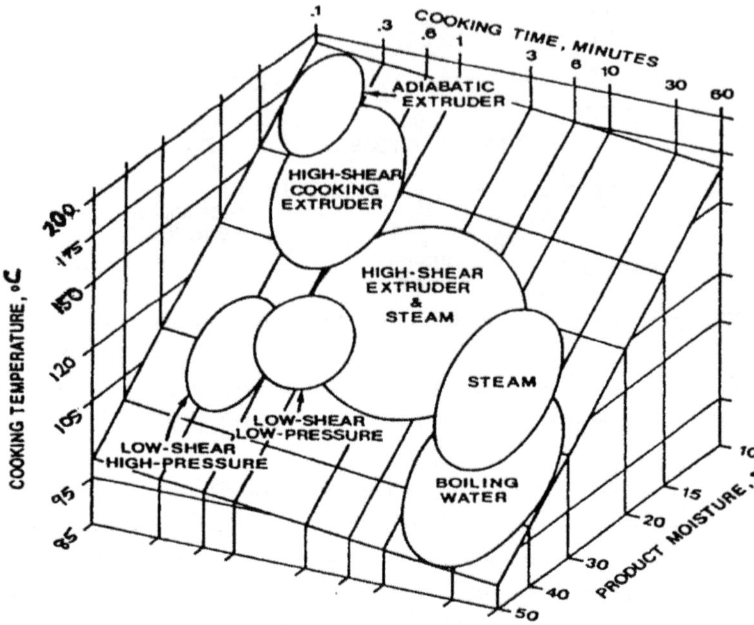

Figure 3 Range of cereal cooking conditions (time, temperature and moisture), with typical ranges of various cooking process cycles (Ref. 3).

- Pasting cooking where both the thermal energy input and the water content are maximized. This process, often described as occurring in excess water, is the 'true' gelatinization process. A typical example of this type of process is the preparation of whole grains for the manufacture of shredded breakfast cereals.

As shown in Fig. 4, another defining characteristic for the various pathways is the shear rate associated with the given process. This is defined via the ratio of the velocity differential/gap (5). The impact of the shear force upon starch components of product formulation is shown in Fig. 5. High shear rates (usually associated with high mechanical energy input processes) promote dextrinization or stress fragmentation (6). The name stems from the fact that numerous low molecular weight carbohydrates (or dextrins) are generated from the starch in the formulation as a result of the shearing process.

Each of the pathways will yield a different free polymeric matrix, as shown in Fig. 3. The composition of the matrix (with respect to the mean molecular weight distribution (MMWD)) will differ, depending upon the chosen pathway. Some of the relative differences between various pathways are shown in Fig. 6. In summary, the major effect of an increase in SME input during processing is that the mean molecular weight of the starch component of the product will be

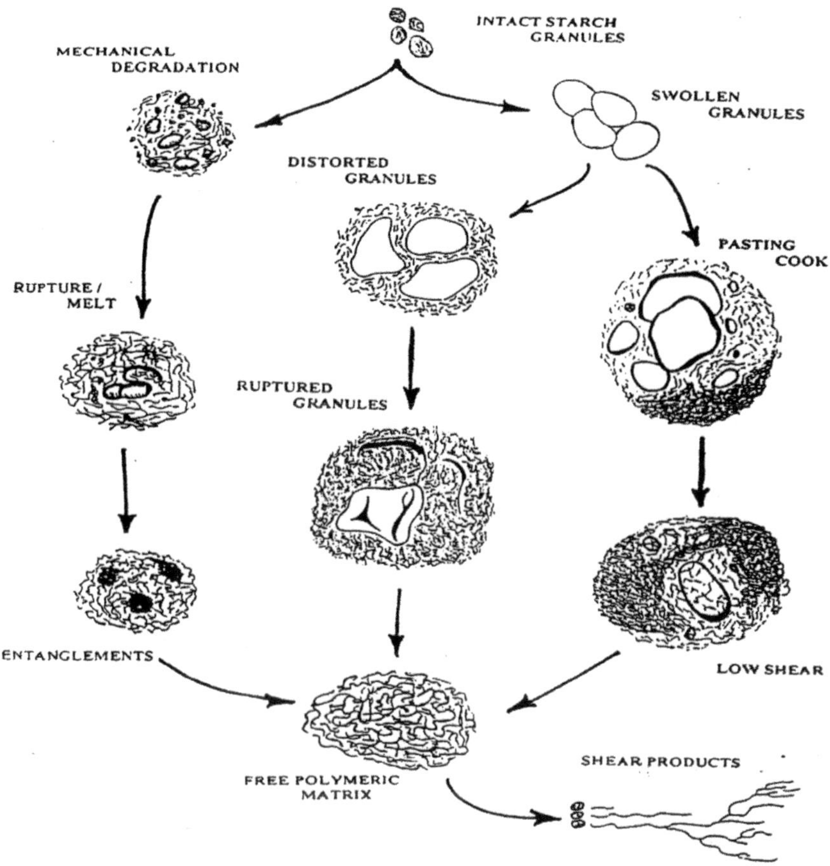

Figure 4 Alternative pathways for starch degradation as affected by in-process moisture content.

reduced, as a result of the effect of dextrinization on the starch components. It is also likely that the same mechanism (i.e. shear) may also be responsible for the breakdown of the protein components of the formulation into their peptide/amino acid building blocks.

4 Developing animal feed products: a case study

The nutritional value of a given product depends on the ingredients in the formulation, the processing parameters used to manufacture the product and on the digestive characteristics of the animal species consuming the feed product. Each animal species has its own specific gastric anatomy. The example of the domestic cat is shown in Fig. 7(a). This highly complex anatomy may be

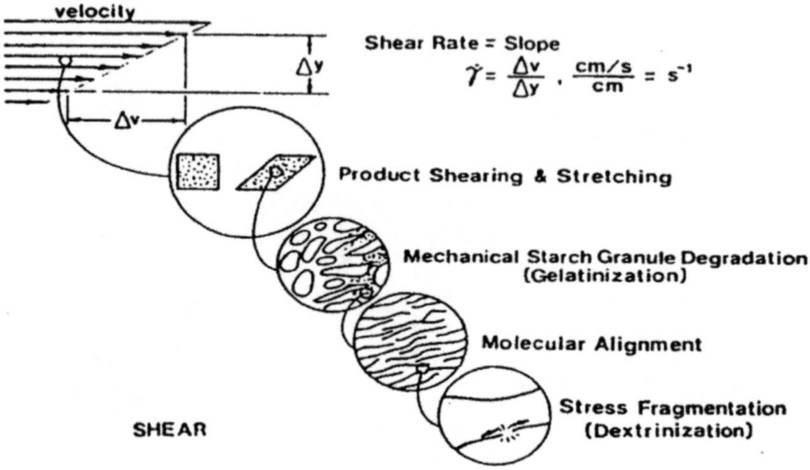

Figure 5 The impact of shear rate on starch.

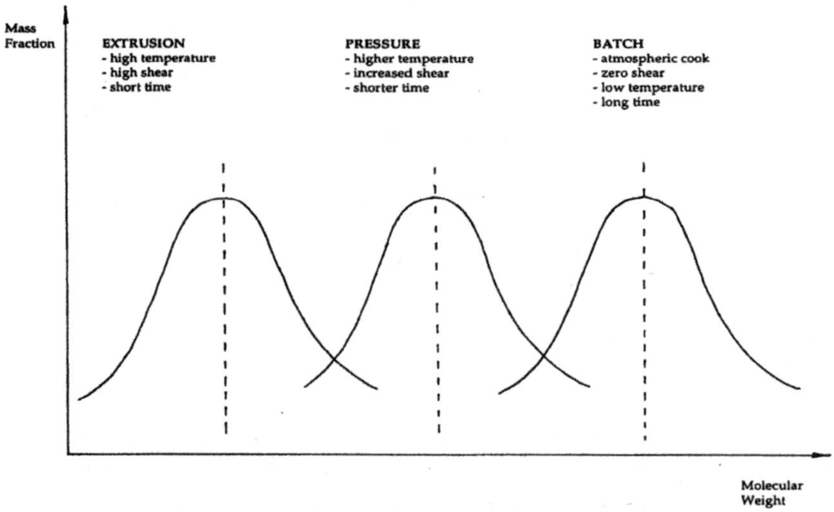

Figure 6 Molecular weight distribution (MWD) of starch as a function of the degradation pathway.

simplified into an engineering model of a gastric system, shown in Fig. 7(b). The efficiency of nutrient extraction (i.e. the nutritional value of the food) will be directly influenced by the viscosity of the digesta which will, in turn, be affected by the rheological properties of the starch/protein components in the food product.

A summary of the gross composition of a range of typical dry pet food formulations currently available on the market is shown in Table 1 (7). From

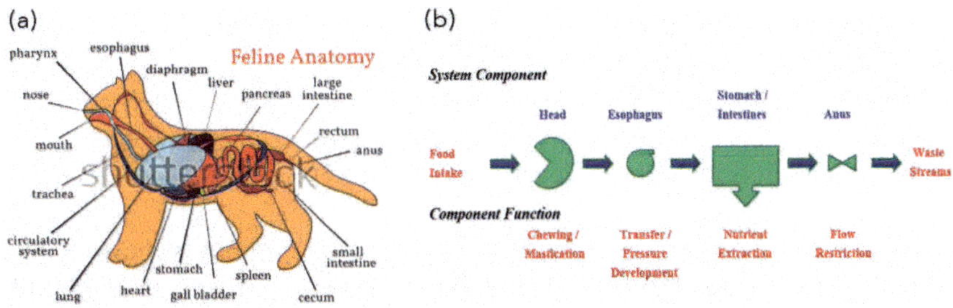

Figure 7 (a) feline gastric anatomy and the (b) engineering model of the gastric system.

this data it is clear that carbohydrates (mostly starch) are the major component of formulations of both dry cat and dry dog food formulations. It can be hypothesised that the effect of extrusion cooking on the starch component will have a significant effect on the quality of the dry cat food product.

In order to show the relationship between the energy input and the starch composition (with respect to MMW), a basic dry cat food formulation (approximating to the composition shown in Table 1) was extruded under a number of different processing conditions. The total formulation (with respect to total moisture and fat addition into the extruder) was maintained for all samples. Adjusting the level of steam injection, the barrel temperatures (i.e. the thermal energy inputs) and/or the extruder screw speed were used to control the specific mechanical energy input for each product sample.

To start with, specific mechanical energy (SME) can be defined as follows:

$$\mathbf{SME} = \left[\mathbf{P}_{\mathbf{operating}} - \mathbf{P}_{\mathbf{no\ load}}\right] / \left[\mathbf{M}\right] \quad \left[\mathrm{kW\,hr\,kg^{-1}}\right]$$

where,

$\mathbf{P}_{\mathbf{operating}}$ = actual operating load of the extruder
$\mathbf{P}_{\mathbf{no\ load}}$ = operating load for the 'empty' extruder
M = total melt flowrate (i.e. dry feed + water + steam + oil + ...)

Table 1 Nutritional targets for dry pet foods

	Dry cat food	Dry dog food
Protein	27	18
Fat	8	12
Moisture	8	8
Carbohydrate	45	51
Fibre	4	4
Ash	8	7

Samples were each analysed for the following three parameters:

- Mean molecular weight of starch (using HPLC analysis);
- Water absorption index (AACC Method 88 - 10); and
- Water solubility index (AACC Method 1632).

The effect of the SME upon the MMW of the starch component is shown in Fig. 8. As anticipated, it was found that there was a significant reduction in the MMW above an SME of approximately 0.015 kW hr kg^{-1}. When the SME exceeded approximately 0.035 kW.hr kg^{-1}, however, the effect appeared to diminish.

This effect may be directly attributed to the increased magnitude of dextrinization (or generation of low molecular weight sugars) which results from the higher level of SME. The formation of dextrins leads to a reduction in the gel-forming capability of the starch, resulting in a reduction in the water absorption index (the WAI is often used as a means of quantifying the water stability of a product). This is shown in Fig. 9. The most significant changes to the WAI are seen to occur within the same range of SMEs outlined above.

Dextrins are known to have significantly higher water solubility, resulting in an increase in water solubility during processing. The water solubility of a product may be readily quantified by determining the water solubility index (WSI). The data for the test samples is shown in Fig. 10. Once again, the most significant changes to the WSI are seen to occur within the same range of SMEs outlined above.

In order to assess the effect of these observed changes in the starch component on the 'eating quality' of the feed, the starch dry matter digestibility

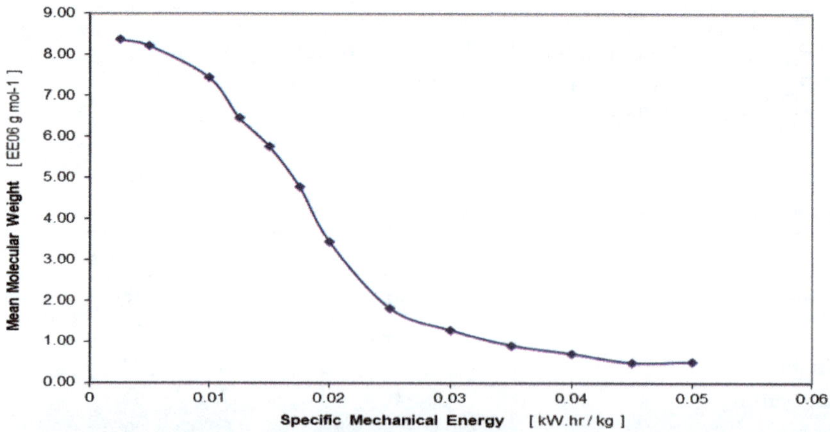

Figure 8 Influence of the specific mechanical energy upon the mean molecular weight (MMW) for extruded wheat-based dry cat food formulation.

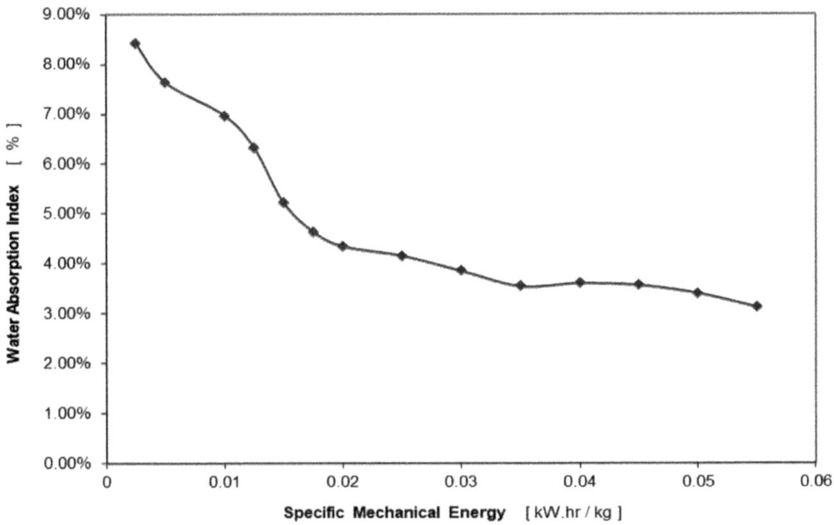

Figure 9 Influence of the specific mechanical energy upon the water absorption index (WAI) for extruded wheat-based dry cat food formulation.

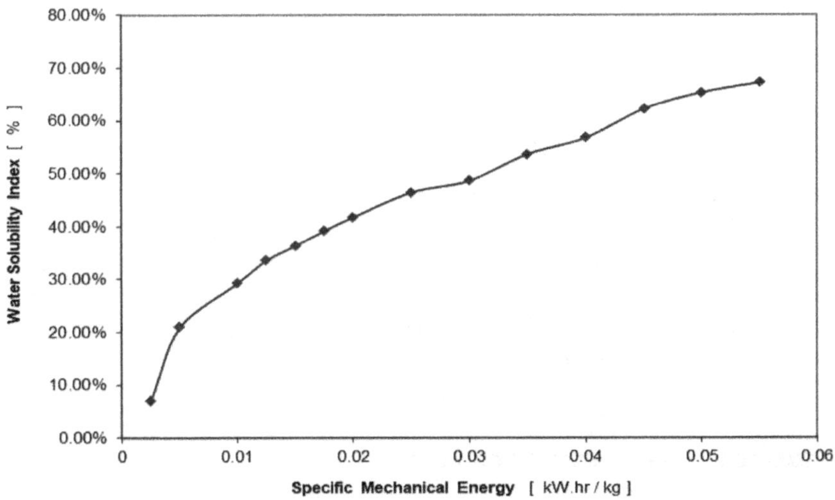

Figure 10 Influence of specific mechanical energy upon the water solubility index (WSI) of extruded wheat starch.

of a commercial dry cat food, prepared under controlled conditions of SME input, was measured. The data was reported on a dry matter basis and the results are presented in Fig. 11. The SDMD was assessed *in vivo* by feeding the products to cats in a commercial cattery under controlled conditions.

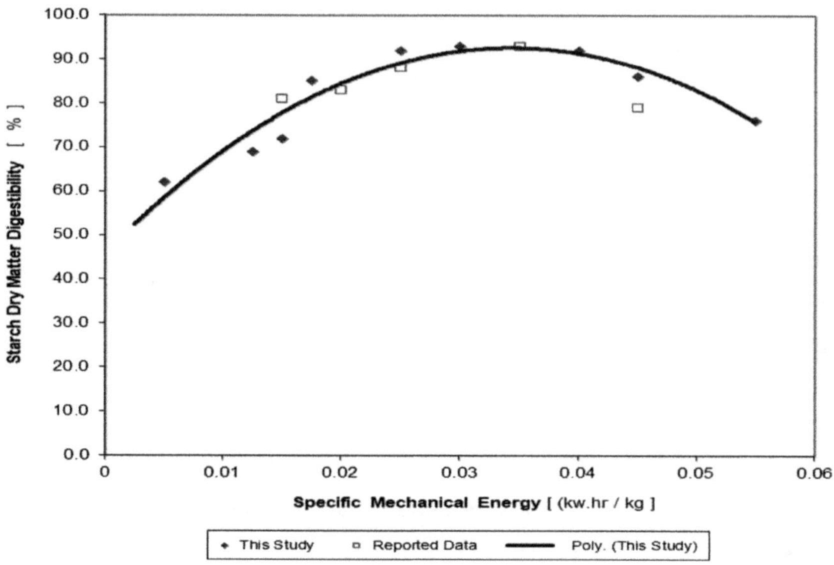

Figure 11 Influence of the specific mechanical energy upon the dry matter digestibility (DMD) of dry cat food.

The data clearly shows the influence of the SME on the starch dry matter digestibility of the food. There would appear to be an optimal range of SME for the manufacture of dry cat food via extrusion cooking, namely

$$0.025\,\text{kW.hr}\,\text{kg}^{-1} < \mathbf{SME} < 0.040\,\text{kW.hr}\,\text{kg}^{-1}$$

This result may be quite easily rationalized since, as the SME is increased, the MMW of the starch components is reduced (see Fig. 8), thereby increasing water solubility (see Fig. 9) and reducing the water absorption capability of the food (see Fig. 10). These changes lead to significant changes in the magnitude of the effective viscosity of the food within the gut. The digesta viscosity will directly influence the transit time through the gut and this will therefore have a direct influence upon the digestibility of the food itself (see Fig. 11).

5 Applications to livestock feed

Throughout the feed industry the feed conversion ratio (FCR) is used to assess animal feed performance and is described as 'kg of feed given to the animal per kg of animal weight gained'. For example, if a flock of poultry birds is given 500 kg of feed throughout the growing cycle and 200 kg of live weight gain is recorded, then the FCR for that flock would be 2.50 (500/200).

Scheideler (8) noted the effect of pellet quality (as measured via the pellet durability index (PDI)) on the FCR for several poultry species, including broilers, roasters and turkeys. Poultry fed pellets with a (PDI) of 70% required less feed to reach market weight than those fed pellets with a PDI of 25% (see Table 2).

Lewis (9) investigated the influence of the process parameters of the hot pelleting process (incorporating both a preconditioner and a pellet press) on 'gelatinization' (i.e. the starch cook, see Table 3) and its impact upon pig growth performance. He concluded that 'the difference in starch gelatinization between hot mash and hot pellet samples suggests that starch gelatinization does occur due to the frictional or mechanical heat of the pelleting process'. (It must be noted, however, that this work was completed at 'very low' moisture content (i.e. $\boldsymbol{w}_M < 25\%$ w/w)).

In summary this research found that only the conditioning temperature, and not the process time, affected available starch in the diet. There were no effects of conditioning temperature or time on vitamin retention. The study concluded 'increasing conditioning temperature and time increased gelatinized starch, but not to the extent of physical pelleting' (i.e. the frictional heating or SME was significant).

These results are in agreement with Wang (10) who showed that mechanical energy was very effective at increasing starch availability during extrusion processing at 50°C. It is also in agreement with Skoch et al. (11) who demonstrated that, when cold mash was pelleted without the addition of steam, 'a significant amount of starch damage was observed due to frictional heat created as feed passed through the die holes through feed compression'. It appears therefore that the formation of pellets results in increased starch availability, rather than increases from temperature or conditioning.

Rokey (12) investigated the benefits of the use of preconditioning with an extended residence time in feed milling (comparing the use of the Wenger DDC (with $\mathbf{t}_R \approx 240$s) versus the traditional single shaft preconditioner (with $\mathbf{t}_R \approx 60$s)). He noted that 'retention time is critical in achieving optimum

Table 2 Effect of pellet quality (as measured via PDI) on the feed FCR

Bird type	PDI = 25%	PDI = 70%
Broilers	2.15	2.08
Roasters	2.36	2.10
Turkeys	2.71	2.61

Table 3 Effects of conditioning temperature and time on starch cook

Conditioning time [s]	15	15	15	30	60
Temperature [°C]	77	88	35.9	35.6	35.5
Total starch [% w/w]	35.2	36.1	35.9	35.6	35.5
% Gelatinized [%]	9.0	10.7	9.6	9.7	10.2

Table 4 Comparison of the effect of extended retention time on pellet quality

	DOC	PDI	FCR
Single shaft cylinder	19.1	35.7	2.13
DDC preconditioner	23.5	48.1	2.11
% Increase	23%	35%	−1%

preconditioning results. DDC technology shortens this time due to excellent mixing properties and a longer retention time means a higher throughput and capacity.'

The degree of cook (DOC) and the pellet durability index (PDI) for pellets prepared with both technologies were measured (via enzymatic digestion and the tumbling can, respectively) and then compared. The results are shown in Table 4. Feeding trials indicated that the FCR for the broiler-fed pellets produced using the DDC was lower than that of the broiler-fed pellets produced on the Single Shaft Preconditioner. The longer retention time and more efficient mixing allow a higher quality pellet to be produced. This can lead to many advantages for feed mills through lower feed to gain ratios and increased PDI.

6 Summary

For many processed food and feed products, the water solubility of the finished product is a very important performance criterion. In some cases, such as ready-to-eat breakfast cereals and dry pet foods, excessive water solubility is to be avoided whilst, in the case of infant feed formulations and snack foods, increased water solubility (i.e. hydration or wettability) is deemed to be more beneficial for product performance. Extrusion cooking technology is a very common processing technology used in the manufacture of a broad range of food and feed products. It is important to have a clear understanding of the mechanisms involved in the determination of the water solubility of extrusion cooked products.

It has been shown that the conversion (or cooking) of the starch component of a food formulation may proceed via any of a number of possible routes (or process pathways). The moisture content of the system during processing is one of the key determinants that dictate the level of specific mechanical energy (or SME) to the process. It is the magnitude of this SME (and other energy inputs) that ultimately dictates the level of degradation of the starch (and protein components) that will occur.

The analysis of an extrusion cooked dry cat food formulation shows that there is a clear relationship between the mean molecular weight (of the starch component), water stability (as measured via the WAI), water solubility (as measured via the WSI) and the SME used in the manufacture of the product samples. Samples of a commercial dry cat food formulation prepared at a range of processing conditions have been shown to generate a range of degrees of starch dry matter digestibility, depending on the magnitude of the SME inputs.

7 References

1. Harper, J. M. 1993. A comparative analysis of single and twin screw extruders, Presented at the AACC Short Course on Food Extrusion, Kansas City.
2. Shahidi, F. et al. 1988. *Process-induced Chemical Changes in Food*, Springer, New York.
3. Miller, R. C. 1994. *Breakfast cereal extrusion technology*. In Frame, N. (ed.), The Technology of Extrusion Cooking, Springer, Dordrecht, The Netherlands.
4. Cooking pathways, *Personal Communication with P. Whalen, Whalen Consulting* 1997.
5. Forte, D. and Young, G. 2016. *Food and Feed Extrusion Technology – An Applied Approach to Extrusion Theory*. Food Industry Engineering. ISBN 978-0-9945433-0-1.
6. Hardacre, A. K., et al. 2018. Predicting the viscosity of digesta from the physical characteristics of particle suspensions using existing rheological models, *Journal of Royal Society* 20180092.
7. *Feeding the Dog and Cat*. 2005. Waltham Centre for Pet Nutrition.
8. Scheideler, S. E., Beck, M. M., Abudabos, A. and Wyatt, C. L. 2005. Multiple-enzyme (Avizyme) supplementation of Corn-Soy-based layer diets, *Journal of Applied Poultry Research* 14(1):77–86.
9. Lewis, L. L. 2011. Evaluation of pelleting process parameters on feed nutrients, starch gelatinization and pig growth performance, MSc Thesis, Kansas State University.
10. Wang, S. S. 1993. Gelatinization and melting of starch and tribo-chemistry in extrusion, *Starch* 45(11):388–390.
11. Skoch, E. R., Behnke, K. C., Deyoe, C. W. and Binder, S. F. 1981. The effect of steam conditioning rate on the pelleting process, *Animal Feed Science and Technology* 6(1):83–90.
12. Rokey, G. Better pellets with DDC preconditioning. *Feed Tech* 6(3):15. www .AgriWorld.nl.

Chapter 3

Processing techniques to optimize digestibility and nutritional value of animal feed

Charles Stark, Kansas State University, USA

1 Introduction

The feed manufacturing process involves a number of steps that convert individual ingredients with different physical characteristics and nutrient contents into a complete feed for livestock and poultry. The feed manufacturing process can be as simple as grinding, weighing and mixing of ingredients to create a mash/meal feed. Alternatively, it can be as complex as grinding, weighing, mixing, long-term conditioning, pelleting and post-pellet liquid application. Whole grains are typically ground at the feed mill, whereas ingredients such as soybean meal, rendered products, distillers' dried grains and wheat by-products are ground by the supplier. The individual ingredients are then weighed into a batch and then mixed where liquids can be applied. A feed mill equipped with a pellet mill can then convert the mash to a pellet and then routes it to bulk load out or packaging. The majority of the feed manufactured for integrated swine and poultry is shipped in bulk, whereas commercial feed mills ship both bulk and bagged feed.

http://dx.doi.org/10.19103/AS.2020.0083.09

The greatest improvement in nutrient digestion occurs during particle size reduction. While poultry and swine have different particle size requirements, the fact still remains that digestibility and consequently animal performance is improved when the particle size of cereal grains is decreased. The effect of particle size reduction has been well documented in cereal grains. The improvement in digestibility of other ingredients such as bran, SBM and DDGs is not as consistent in the literature. The particle size of by-products is often reduced to improve pellet quality.

Owens and Heimann (1994) indicated that the major reasons for particle size reduction are to:

- expose greater surface for digestion;
- improve ease of handling of some ingredients;
- improve mixing characteristics of ingredients; and
- improve pelleting efficiency and pellet quality.

While all are important, exposing the endosperm and increasing the surface area of the starch has the greatest effect on improved starch digestibility. The target size of the particle of the ground grain will vary based on age and species of the animal. Swine require a much smaller particle size than do poultry. The target particle size for swine is between 400 microns and 600 microns while the range for poultry is from 600 microns to 1000 microns for broilers and from 1000 microns to 1200 microns for layers. While research has been conducted in this area for more than 40 years, the absolute particle size targets for livestock and poultry are not well defined due to variability in the particle analysis procedure. Most researchers only report the particle size of ground grains and do not report the particle size of the major ingredients in the diet or the method used to obtain the particle size, which creates a challenge when comparing the results of different studies (Stark, 2017). Therefore, producers, nutritionists and researchers should only evaluate the trends in the data unless the report outlines all procedures used to determine the particle size.

The cost of particle size reduction is dependent on the target particle size, the type of equipment (hammermill versus roller mill), and the ingredients that must be processed (Stark, 2012). The cost is associated primarily with electricity, screens, hammers and re-corrugation of rolls. The cost of grinding is inversely related to the particle size; the cost will increase as the target particle size decreases. Electricity makes up the highest percentage of the grinding cost, followed by maintenance and equipment depreciation. In some cases, the cost of electricity could be ten times the cost of maintenance, and the electricity cost will increase as the parts (hammers, screens and roll) of the equipment continue to wear. At some point in the maintenance cycle the cost of not replacing the parts will drive up the total grinding cost. The loss in animal performance due to poorer feed conversions will exceed the cost of maintenance in most feed mills. Goodband et al. (2002) estimated that an increase of 100 microns can increase feed conversion by 1% in swine, and can significantly affect the profitability of a company.

The benefits of pelleting mash feed are to improve the flowability of the feed, reduce feed wastage and improve the palatability of the feed. While some researchers and nutritionists believe there is an improvement in starch digestion due to gelatinization of the starch, there is little data to support the theory. The research does show an improvement in feed conversion when animals receive pelleted feed, but it is often confounded with particle size in both poultry and swine.

2 Methods of particle size reduction: hammermills and roller mills

The two most popular methods for particle size reduction are:

- the hammermill; and
- the roller mill.

The advantages and disadvantages of each piece of equipment must be considered when selecting the type of equipment that is used in the feed mill. The type of equipment should be based on the type of ingredients that will be used in the diet, target particle size, manufacturing capacity, number of employees, operational costs (electricity and maintenance) and the final form of the feed (meal vs. pellet). Corn and sorghum can efficiently be ground to less than 500 microns with a roller mill but grinding higher fiber cereal grains such as barley, triticale and wheat may result in greater variation and separation of the bran and endosperm components as the material moves through the feed mill. Additionally, a double pair roller mill can be used to efficiently grind cereal grains to a particle size of 500 microns. However, if the target particle size is below 500 microns, a triple or quad roller mill must be used to consistently grind grain to the target particle size. A hammermill is a universal method that can reduce the particle size of high fiber grains and ingredients to less than 400 microns when properly sized and fitted with the correct air assist system.

2.1 Hammermills

Particle size reduction in a hammermill occurs as a result of the impaction between the rapidly moving hammers and a relatively slow-moving particle (Heiman, 2005). The impaction continues to occur until the velocity of the particle has decreased and is slow enough to exit through the opening in the screen. Changes in moisture, fiber content and genetic variety of ingredients

Table 1 Effect of hammer tip speed and screen size opening on the particle size (microns) of US #2 corn

Tip Speed (m/min)	Screen Size, mm				
	1.6	2.4	3.2	4.0	4.8
	d_{gw}, μm				
5455	450	600	750	950	1250
6317	300	500	650	750	900
7752	200	400	550	650	750

tend to have less of an effect on the particle size when ingredients are ground with a hammermill as compared to a roller mill. The hammermill will produce a product with a wider variation in particle diameters, especially when equipped with a large screen. The tip speed, number of hammers, hammer setting (coarse vs. fine) and screen size affect the particle size of grain ground with a hammermill. However, tip speed and screen size will have the greatest impact on the particle size. Heiman (2005) reported that increasing the tip speed of the hammer will produce a smaller particle size within a given screen (Table 1).

Evans et al. (2018) reported an interaction between screen hole diameter (2 mm, 4 mm, and 6 mm.) and hammer tip speed (3124 m/min, 4686 m/min, and 6248 m/min) when grinding whole corn. The results indicated an interaction that when the tip speed increased from 3124 m/min to 6248 m/min, the rate of decrease in d_{gw} was greater as the screen hole diameter increased from 2 mm to 6 mm. Therefore, when the tip speed was increased from 3124 m/min to 6248 m/min, d_{gw} was reduced by 233, 258 and 305 microns for corn ground using the screen hole diameter of 2 mm, 4 mm and 6 mm, respectively.

2.2 Roller mills

The roller mill will produce a granular product by passing the product through a pair of horizontally mounted steel rolls. The particle size is controlled by the width of the gap between the rolls, number of corrugations per inch on the roll, the spiral of the corrugation, and the differential speed of two rolls within the pair. The configuration of roller mills has changed over the last 10 years as suppliers have introduced both three and four pair roll stands as compared to a two pair roll stand that has been used for the last 50 years. The introduction of newer technology has allowed feed manufacturers to produce a smaller, more uniform grind. In the past, roller mills could only consistently achieve a particle size of 500 microns. Gebhardt et al. (2018) reported particle sizes of 525, 394 and 267 microns using 2, 3 and 4 pair roller mills, respectively. Groesbeck et al. (2003) reported that roller milled corn samples had a narrower range of particle diameters (S_{gw} 1.83–2.03) in samples whose particle size (d_{gw})

ranged from 503 microns to 1235 microns. In contrast, the range of particles in hammermill ground corn was wider (S_{gw} 2.12–2.52) in the samples, ranging from 390 microns to 980 microns, and the reported variation decreased as the particle size of the corn decreased. The primary advantage of the roller mill as compared to a hammermill is that it processes 15–40% more tons/h to a similar particle size using the same horsepower (Heiman, 2005) when the target particle size is above 500 microns. The throughput of a roller mill is significantly reduced when the target particle size is below 500 microns, which results in a similar cost to grind the grain. Roller mills tend to require more labor to monitor the particle size, adjust the roll gap settings and change out the rolls.

3 Particle size analysis

The approved method for particle size analysis is ASAE S319.4 "Method of Determining and Expressing Fineness of Feed Materials by Sieving" (ASABE, 2012). However, the method lacks specific details that can significantly change the results of the ground material. Fahrenholz et al. (2010) first evaluated different pieces of testing equipment, methods, and times (Table 2). The data indicated that sieve shaker, sieve agitators, dispersion agent, and sieving times influenced mean particle size, or geometric mean diameter (d_{gw}), and variation (S_{gw}). Stark and Chewning (2012) first demonstrated that the particle size of a sample can vary by over 150 microns based on how the test is performed (Table 3).

Kalivoda et al. (2017) evaluated technician, mill type, sieving agent, sieve agitators and time (Table 4). The results of their research suggested

Table 2 Effect of equipment, method, and time on the mean particle size and standard deviation

Method	Mean, D_{gw} (microns)	Standard Deviation, S_{gw}
Sieve Shaker		
Tyler Ro-Tap	589	2.11
Retsch	497	2.53
Sieve Agitators		
With	523	2.4
Without	624	2.0
Dispersing Agent		
With	486	2.46
Without	560	2.10
Sieving Time		
10 minutes	523	2.40
15 minutes	481	2.56

Table 3 Mean particle size and standard deviation of corn ground through different hammermill screens and analyzed with different methods

Method	Mean, D_{gw} (microns)	Standard Deviation, S_{gw}
Fine (2.4 mm)[1]		
Without	554	1.64
Agitators	478	1.82
Dispersing Agent	329	2.56
Medium (4.8 mm)[1]		
Without	659	1.89
Agitators	610	2.05
Dispersing Agent	407	3.15
Coarse (9.6 mm)[1]		
Without	886	2.42
Agitators	832	2.53
Dispersing Agent	648	3.63

[1] Hammermill screen hole openings.

Table 4 Effects of method × grain type on geometric mean diameter (d_{gw}) and geometric standard deviation (S_{gw})

	Method				
Sieving time, min	10	10	15	15	15
Sieve agitator inclusion	Yes	Yes	No	Yes	Yes
Dispersion agent inclusion	No	Yes	No	No	Yes
Geometric mean diameter (d_{gw}), μm					
Corn	544	498	586	530	486
Wheat	656	640	682	647	623
Sorghum	559	524	577	551	512

that the test should be performed with sieve agitators and sieving agent for 10–15 min. The experiment also recommended the type, size, number and location of the sieve agitator. The updated recommendations were published in the Kansas State University Extension bulletin 'Evaluating Particle Size of Feed Ingredients'. A review of the literature highlighted the fact that many researchers do not report the particle size of the grain used in research projects or the particle size methodology. Modifications to the ASAE S319 method, such as the addition of sieve agitators, sieving agents, sample sizes or sieve sizes, must be included in the research methods. Otherwise, the conclusions reported in the research may not be based on the actual particle size of the ingredient.

4 Effect of particle size reduction on swine performance

The benefit of particle size reduction is different in swine and poultry due to differences in the autonomy of the digestive tract. Digestibility of grains is improved in swine when the particle size of the grain is below 600 microns. The research over the last 40 years in swine clearly indicates that there is an improvement in feed conversion as the particle size of the grain is reduced. However, nutritionists must determine the optimal particle size based on the diet that is being fed to the animals and this will change based on least cost formulation. The return on investment is greater for grinding to a smaller particle size when the cost of inputs increases. Therefore, nutritionists, feed mill managers, veterinarians and production specialists must continually evaluate the cost benefits of particle size reduction.

The benefit of particle size reduction of cereal grains for swine has been well documented in several swine studies (Table 5). However, the positive effects of particle size reduction on feed conversion in swine must be balanced between feed mill capacity and any potential adverse effects on animal health, especially in swine that may have a genetic predisposition to gastric ulcers. The majority of the research on the effect of particle size in swine was conducted 30–40 years ago at Kansas State University. However, with the changes in swine genetics and feed manufacturing equipment there has been a renewed interest in particle size reduction research. The particle size that can be fed to swine is also dependent on the physical form of the feed and the percentage of the ground corn in the feed. In corn-SBM mash diets, reducing the particle size below 500 microns results in poorer flowability of the complete feed in feed bins and feeders, which contributes to feed outages. However, in pelleted diets the particle size of the corn can be reduced to less than 300 microns. Producers must evaluate the return on investment to grind to less than 500 microns and determine if the added cost to pellet the diet can be offset by improved feed conversion.

Ohh et al. (1983) demonstrated that fine grinding corn and sorghum improved feed conversion in starter pigs. Healy et al. (1994) reported improvements in starter pig performance when corn was reduced from 1000 microns to 500 microns. However, Healy et al. (1994) observed a negative effect on stomach mucosa at 300 microns. Mavromichalis et al. (2000) reported improved feed conversion in starter pigs when the particle size of wheat was reduced from 1380 microns to 387 microns. However, De Jong et al. (2012) reported that using 620 microns or 352 microns corn in meal nursery diets reduced ADG but did not affect feed conversion of nursery pigs. In a follow-up study De Jong et al. (2013) reported that reducing the particle size of corn from 737 microns to 324 microns improved nursery performance when fed in the pelleted form.

Table 5 Effect of cereal grain particle size reduction on swine performance

Reference	Microns	ADG, kg	F/G	Research Notes
Ohh et al., 1983	624	0.46	1.70	Corn, Starter pigs
	877	0.45	1.78	Corn, Starter pigs
	822	0.46	1.81	Corn, Starter pigs
	1147	0.47	1.92	Corn, Starter pigs
	539	0.44	1.78	Sorghum, Starter pigs
	722	0.45	1.79	Sorghum, Starter pigs
	885	0.45	1.92	Sorghum, Starter pigs
	1217	0.43	1.94	Sorghum, Starter pigs
Healy et al., 1994	369	0.42	1.44	Corn, Starter pigs
	487	0.44	1.41	Corn, Starter pigs
	702	0.41	1.46	Corn, Starter pigs
	919	0.43	1.51	Corn, Starter pigs
Goodband and Hines, 1987	698	0.93	3.39	Sorghum, Finishing pigs
	714	0.89	3.32	Barley, Finishing pigs
	902	0.92	3.56	Barley, Finishing pigs
	1126	0.81	3.65	Barley, Finishing pigs
	2200	0.79	3.72	Barley, Finishing pigs
Wondra et al., 1995	400	0.98/0.99	3.22/3.01	Corn, Finishing pigs
	600	0.95/1.02	3.43/3.13	Corn, Finishing pigs
	800	0.94/1.01	3.41/3.15	Corn, Finishing pigs
	1000	0.96/0.99	3.39/3.32	Corn, Finishing pigs
Mavromichalis et al., 2000	387	0.43	1.19	Wheat, Starter pigs
	568	0.45	1.16	Wheat, Starter pigs
	1380	0.41	1.34	Wheat, Starter pigs
	628	0.89	3.44	Wheat, Finishing pigs
	1288	0.88	3.57	Wheat, Finishing pigs
	406	0.91	3.12	Wheat, Finishing pigs
	614	0.90	3.23	Wheat, Finishing pigs
Paulk et al., 2015	319	1.10	2.65	Sorghum, Finishing pigs
	573	1.10	2.71	Sorghum, Finishing pigs
	724	1.13	2.74	Sorghum, Finishing pigs
De Jong et al., 2016	326	0.93	2.72	Wheat, Finishing pigs
	579	0.91	2.78	Wheat, Finishing pigs
	728	0.92	2.72	Wheat, Finishing pigs
Gebhardt et al., 2018	285	0.94	2.82	Corn, Finishing pigs
	473	0.97	2.81	Corn, Finishing pigs
	561	0.99	2.84	Corn, Finishing pigs

Wondra et al. (1995) reported that reducing the particle size of corn from 1000 microns to 400 microns resulted in an 8% improvement in feed/gain. Mavromichalis et al. (2000) did not detect any change in finishing pig performance by reducing the particle size of wheat from 1288 microns to 628 microns; however, there was a 3% improvement in feed conversion when the particle size of wheat was reduced from 614 microns to 406 microns. Paulk et al. (2015) reported that decreasing the particle size of sorghum improved the feed conversion of finishing swine. While the trends in improvement were similar, the particle size of the sorghum had to be approximately 100 microns lower than that of the corn to achieve similar performance. De Jong et al. (2016) also evaluated the effect of reducing the particle size of different hard red winter and soft white wheat varieties ground to 693, 465, 245 and 710, 402, 258 microns, respectively, and fed in meal diets. There was an overall linear improvement in G:F (0.354, 0.361, 0.367) and DM digestibility (89.0%, 91.5%, 91.5%) as the particle size of the wheat was reduced from 728 microns to 325 microns. However, in a subsequent study no effect on performance was observed when similar particle sizes of hard red and soft white winter wheat were fed in the form of pellets. Acosta et al. (2019) reported a difference in apparent total track digestibility of corn processed with a hammermill and a roller mill. The research suggested that corn ground to a smaller particle size with the roller mill had improved digestibility as compared to a similar size with a hammermill. Goodband et al. (2002) developed a regression equation to predict feed conversion from 400 microns to 1200 microns in finishing pigs from 55 kg to 110 kg, suggesting approximately a 1% improvement in feed conversion for every 100-micron reduction in particle size.

5 Effects of particle size reduction on broiler performance

Research has shown that increasing the particle size of cereal grains in broiler diets has improved broiler performance, improved digestibility, reduced nitrogen excretions and improved gut health. Studies showed there is a correlation between the particle size of corn and gizzard development and feed utilization. The gizzard, along with reverse peristalsis within the GI tract, improves nutrient digestibility and gut health (Ferket, 2000). Larger gizzards relative to BW have resulted in improved feed utilization (Xu et al., 2015a) and gastric intestinal tract health (Ferket, 2000). Nir et al. (1994) stated that coarser feed particles increased relative gizzard weight while Amerah et al. (2008) suggested gizzard stimulation was due to the length of time that the coarse particles resided in the gizzard. Differences in research results may be influenced by whether the birds were raised in cages or floor pens as well as the quality of the litter within the pens. Birds that have access to litter may consume the litter in an effort to stimulate the gizzard when coarse particles of grains are not provided in the diet.

Several studies conducted with broilers and turkeys have not shown any benefit in reducing the particle size of corn and wheat (Table 6). Deaton et al. (1995) reported no difference in boilers fed corn ground through 3.2 mm, 6.4 mm, and 9.5 mm screens. Svihus et al. (2004) and Amerah et al. (2007) reported benefits to feeding wheat-based meal diets that included larger particles. However, the advantage of large particles in the feed tends to diminish when the diets are fed as pellets. Amerah et al. (2007) reported coarseness of grind affected growth and feed conversion in birds fed mash, but no difference was observed within birds fed pellets. Martinez and Parsons (2007) reported no advantage in phosphorus availability or bird performance by reducing the particle size of DDGS. However, the most recent series of experiments conducted by Xu et al. (2015a,b,c,d, 2017) demonstrated an improvement in broiler performance when coarse particles were incorporated into the pelleted feed. The inclusion of either 25% or 50% coarse corn (0%, 25%, 50%) improved FCR at 35 d (1.82, 1.74, 1.69) and 42 d (1.94, 1.86, 1.82), respectively (Xu et al.,2015a). Furthermore, inclusion of coarse corn also increased digesta retention time by 11% and 27% at d 35 and 42, respectively. The coarse particles also improved litter characteristics due to the improved digestion of nutrients due to improved

Table 6 Effect of particle size reduction on brolier performance

Reference	Microns	Gain,g	F/G	Comments
Deaton et al., 1995		2656/2227	1.90/1.97	3.18 mm screen, 21 C/31 C, Broiler
		2652/2196	1.90/1.97	6.35 mm screen, 21 C/31 C, Broiler
		2651/2217	1.89/1.97	9.53 mm screen, 21 C/31 C, Broiler
Svihus et al., 2004	600	1076/1361	1.72/1.57	Wheat, Hammermill, Mash/Pellet, Broiler
	930	1089/1348	1.70/1.60	Wheat, Hammermill, Mash/Pellet, Broiler
	670	1029/1333	1.68/1.58	Wheat, Roller mill, Mash/Pellet, Broiler
	920	993/1350	1.71/1.57	Wheat, Roller mill, Mash/Pellet, Broiler
	1700	1135/1315	1.64/1.58	Wheat, Roller mill, Mash/Pellet, Broiler
Amerah et al., 2007	839	439/834	1.72/1.53	Wheat, Hammermill, Mash/Pellet, Broiler
	1164	539/824	1.63/1.52	Wheat, Hammermill, Mash/Pellet, Broiler
Xu et al., 2015a		2929	1.94	100% Fine Corn, Pellet, Broiler
		3118	1.86	25% Coarse Corn, Pellet, Broiler
		3059	1.82	50% Coarse Corn, Pellet, Broiler
Xu et al., 2015b		3860	1.92	100% Fine Corn, Pellet, Broiler
		3899	1.88	50% Coarse Corn, Pellet, Broiler

gizzard function. The addition of 50% coarse corn decreased the moisture and nitrogen content of the litter (Xu et al., 2015b,c). However, the coarse ground corn does present a challenge in pelleted diets when a high-quality durable pellet is specified with minimal fines. The nutritionist, veterinarian production manager and feed mill manager must determine the cost-benefit ratio of coarse corn versus pellet fines as achieving both is not economically feasible in most operations.

6 Pelleting feed

The two main reasons for pelleting are to improve feed efficiency and improve feed handling characteristics (Fairfield, 2005). Historically, animal producers have observed a 6–8% improvement in performance when animals are fed pellets. However, the actual affect that pelleting has on performance is not conclusive, as there are a number of confounding factors such as particle size of the grain, pellet quality, and fat addition levels. The literature contains theories on improved digestibility as a result of starch gelatinization during conditioning and pelleting. However, the results do not indicate a significant increase in starch gelatinization during pelleting. Skoch et al. (1981) reported minimal gelatinization during the conditioning process and, in fact, most of the starch gelatinization occurred on the surface of the pellet as the pellet moved through the pellet die due to the frictional heat created as the feed slides against the surface of the metal die. Lewis et al. (2015) reported only a 3–5% increase in starch gelatinization as compared to mash diets after conditioning to 77°C and 88°C (mash 6%, 77°C – 9%, 88°C – 10.7%). Saensukjaroenphon (2019) reported an increase in starch gelatinization in complete feed from 7.2% in the mash to 14.7% in feed pelleted at 79°C.

The pelleting process improves the physical characteristics of feed and makes it more palatable. However, the high temperatures associated with the pelleting process have negative effects on enzyme stability that should be considered by nutritionists. The addition of steam during the conditioning process and friction created in the pellet die can increase the temperature of the mash feed by 60–70°C. Pope (2018) reported a linear decrease in xylanase activity from 100% to 20% as the temperature across the die increased from 0°C to 8°C, respectively. Additionally, he reported a linear decrease in phytase activity as the conditioning temperature increased from 80°C to 92°C, with the most significant losses occurring after the pellet die. Truelock et al. (2019) reported a linear decrease in phytase stability when the temperature and die thickness were increased in both mash and cooled pellets, thus indicating that the pellet die has a significant impact on enzyme stability. Enzyme stability was significantly reduced when the temperature of either the conditioned mash or hot pellets exceeded 80°C. Saensukjaroenphon (2019) reported that, in addition to conditioning temperature and hot pellet temperature, researchers should also consider die retention time. Die retention time is calculated based on the pellet mill throughput, effective thickness of the die and die hole volume. Saensukjaroenphon (2019) reported that increasing the effective thickness:pellet hole diameter (L:D) from 6.6 to 10.6 with a 2.6 sec die retention time resulted in a decrease in phytase stability from 72% to 9%, respectively. However, if the die retention time was 1.6 sec the decrease in phytase stability went from 81% to 75% when the L:D ratios were 6.6 and 9.3, respectively. The results of the research suggest that researchers must measure both conditioned mash and hot pellet temperatures as well as die retention time when conducting feed processing research.

7 Effect of pelleting on swine performance

The effects of pelleting on swine performance have been well documented, especially when ingredients are expensive (Table 7). The performance advantage from feeding pellets must be high enough to cover the added costs of equipment and operating expenses required to produce a pellet. Skoch et al. (1983) reported an 8% improvement in ADG and F/G in nursery pigs. Stark (1994) reported similar benefits to feeding a pelleted diet to nursery pigs, with a trend for better feed conversion when pigs were fed screened pellets versus 25% fines. There was also a tendency for poorer feed conversion when finishing pigs were fed a greater percentage of pellet fines. Pigs fed a pelleted finisher diet that contained 60% fines resulted in a feed conversion similar to feeding mash diets, thus negating the benefit of pelleting. However, research conducted by Hanrahan (1984) did not show an improvement in feed conversion. The improvement in performance may be tied to pellet quality. Harris et al. (1979) reported an improvement by feeding quality pellets versus poor quality pellets in growing-finishing pigs. Schell and van Heugten (1998) also reported poorer feed conversion in grower pigs as the level of fines in the diets increased from 3% to 40%. Poorer quality pellets result in greater feed wastage, especially if the feeders are not properly adjusted to prevent feed wastage. Nemechek et al. (2015) reported that feeding poor quality pellets in the nursery (33% fines) and finishing (50% fines) resulted in poor feed efficiency, which was a result of higher feed disappearance or feed wastage. Nemechek et al. (2015) suggested that reducing the feeder gap from 2.5 cm to 1.3 cm reduced feed disappearance and improved G:F in finishing pigs. The study also demonstrated that as the pigs grew, the pan coverage at 1.3 cm increased to 41%, 70% and 91% at 22, 48 and 91 days, respectively.

Table 7 Effect of pellets on swine performance

Reference	Diet Form	ADG, kg	F/G	Comments
Skoch et al., 1983	Mash	0.61	2.21	Nursery
	Pellet	0.67	2.03	Nursery
Harris et al., 1979	Mash	0.61	3.85	Grow-Finish
	Poor pellets	0.66	3.80	Grow-Finish
	Good pellets	0.66	3.30	Grow-Finish
Hanrahan, 1984	PDI - 69%	0.49	3.99	Grow-Finish
	PDI - 62%	0.49	3.94	Grow-Finish
Stark, 1994	Mash	0.47	1.70	Nursery
	Pellet - 0% fines	0.49	1.50	Nursery
	Pellet - 25% fines	0.49	1.54	Nursery
	Mash	0.43	1.95	Nursery
	Pellet - 0% fines	0.47	1.65	Nursery
	Pellet - 15% fines	0.46	1.72	Nursery
	Pellet - 30% fines	0.45	1.70	Nursery
Stark, 1994	Mash	0.93	2.77	Grow-Finish
	Pellet - 0% fines	0.96	2.65	Grow-Finish
	Pellet - 20% fines	0.96	2.78	Grow-Finish
	Pellet - 40% fines	0.96	2.77	Grow-Finish
	Pellet - 60% fines	0.94	2.82	Grow-Finish
Wondra et al., 1995	Mash	0.96	3.35	Grow-Finish
	Pellet	1.00	3.16	Grow-Finish
Schell and van Heugten, 1998	Pellet - 2.5% fines		1.92	Grower
	Pellet - 13% fines		1.94	Grower
	Pellet - 25% fines		1.98	Grower
	Pellet - 40% fines		2.00	Grower
	Pellet - 3% fines		2.01	Grower
	Pellet - 12% fines		2.08	Grower
	Pellet - 23% fines		2.09	Grower
	Pellet - 37% fines		2.10	Grower
Steidinge et al., 2000	Meal		1.39	Nursery
	Meal 5% SDAP		1.37	Nursery
	Pellet 60C		1.37	Nursery
	Pellet 66C		1.35	Nursery
	Pellet 71C		1.35	Nursery
	Pellet 77C		1.35	Nursery
Nemechek et al., 2015	Meal	0.63	1.49	Nursery
	Poor Pellet	0.59	1.48	Nursery
	Pellet	0.59	1.42	Nursery
	Meal	0.97	2.92	Finish
	Poor Pellet	0.99	2.78	Finish
	Pellet	1.00	2.58	Finish

8 Effect of pelleting on broiler performance

The selection of fast-growing broilers along with higher live weights now requires pelleted feed to achieve the feed intake required to meet the nutrient requirements of the birds. Research studies have consistently reported an improvement in feed intake and conversion (Table 8) when birds were fed pellets versus mash feed (McKinney and Teeter, 2004; Lemme et al., 2006; and Amerah et al., 2007). Chewning et al. (2012) reported improved BW and improved FCR in birds fed pelleted versus mash diets, but also observed that birds fed finely ground particle corn particles (300 versus 600 microns) had

Table 8 Effect of pellets on broiler performance

Reference	Diet Form	Gain, g	F/G	Comments
Moritz et al., 2001	Meal	2089	1.69	Low moisture
	Meal	2043	1.85	High Moisture
	Pellet, PDI - 51%/62%	2226	1.66	Low moisture
	Pellet, PDI - 77%/87%	2186	1.77	High Moisture
Greenwood et al., 2004	Pellet - Fines - 20%	1067	1.51	Broiler
	Pellet - Fines - 30%	1041	1.50	Broiler
	Pellet - Fines - 40%	1033	1.52	Broiler
	Pellet - Fines - 50%	1022	1.52	Broiler
	Pellet - Fines - 60%	983	1.54	Broiler
McKinney and Teeter, 2004	Meal	643	2.02	Broiler
	Pellet - 100%, Fines - 0%	725	1.87	Broiler
	Pellet - 80%, Fines - 20%	701	1.88	Broiler
	Pellet - 60%, Fines - 40%	687	1.92	Broiler
	Pellet - 40%, Fines - 60%	685	1.93	Broiler
	Pellet - 20%, Fines - 80%	675	1.95	Broiler
Parsons et al., 2006	Pellet, Soft - PDI - 90	1604	1.98	Broiler
	Pellet, Hard - PDI - 86	1711	1.90	Broiler
Lemme et al., 2006	Meal - Coarse	1812	1.57	Broiler
	Pellet, PDI - 25%	1822	1.60	Broiler
	Pellet, PDI - 87%	1934	1.63	Broiler
Amerah et al., 2007	Meal	496	1.67	Broiler
	Pellet, PDI - 33%	829	1.52	Broiler
Corzo et al., 2011	Meal	2354	1.78	Broiler
	Pellet - 32% Fines	2556	1.74	Broiler
	Pellet - 64% Fines	2557	1.71	Broiler
Lemons and Moritz, 2016	Pellet - 30% Fines	2433	1.65	Broiler
	Pellet - 60% Fines	2331	1.68	Broiler

improved FCR in the mash versus pelleted form. Furthermore, the relative response to the pelleted form of the diet was greater in faster growing males versus females. Idan (2019) reported that feeding micro pellets (2.4 mm) or high-quality crumbles with the fines removed increased 21 d BW 16% as compared to meal diets. Hu et al. (2012) reported that birds fed crumbles versus mash exhibited a higher feed intake, which resulted in a greater BW when fed in cages. However, most of the positive effect on BW that occurred from feeding high quality crumbles was no longer apparent at 35 d of age after the broilers were moved to the floor pens. The advantage to producing feed with a higher percentage of crumbles and pellets versus fines has been well documented in the literature (Nir et al., 1995; Engberg et al., 2002; Svihus et al., 2004; Corzo et al., 2011; Lemons and Moritz, 2016). Pacheco (2014) reported feed containing up to 25% fines could be fed to broilers without adverse effects on live performance. McKinney and Teeter (2004) suggested there was no advantage to producing more than 40% pellets unless pellet quality was in excess of 60%. Additionally, Lemons and Moritz (2016) reported that feeding a higher percentage of crumbles and pellets as compared to fines resulted in improved broiler performance, especially in pens with a high bird density. Improvements observed due to pelleting could be from the increased nutrient density of the pellets, improved starch and protein digestibility, increased nutrient intake due to the physical form of the feed, reduced feed wastage, and decreased energy expenditure during feeding (Amerah et al., 2007). While the research continues to support improved broiler performance when broilers are fed less fines, the maximum specification of fines at the pan varies by company. The return on investment for producing high quality pellets varies by company depending on housing conditions, genetics, ingredient prices, environmental conditions and feed manufacturing capacity and costs. Optimization of the production system must occur based on the target weight of birds, ingredient costs, production constraints and feed manufacturing costs. Therefore, the economic model used in one company, country or region must be carefully evaluated before applying it to another location in the world.

9 Conclusion

There continues to be research on the effect of feed processing on animal performance worldwide. Changes in animal genetics, feed ingredients, and reduced usage of feed grade antibiotics will require more research in this area. Advances in the field of feed processing research and improved digestibility can be accelerated if more feed processing information is included in the materials and methods of research articles. Differences in particle size analysis methods, pellet quality measurements and the lack of reporting of processing methods make comparisons between animal studies difficult since the feed manufacturing process can create confounding results in the data. In many cases, the results of a treatment will be different in mash and pelleted feeds. Factors that contribute to these differences include the particle size of the grain, pellet quality, percentage of fines in the feed, and palatability of the diet. Researchers, nutritionists and feed manufacturers must take into consideration how much of an effect the feed manufacturing process has on animal performance. The results of experiments can be changed by simply changing a few processing parameters. While the feed manufacturing process is fairly straight forward, the effect that feed manufacturing has on ingredient digestibility is significant and should be considered when evaluating research results.

10 Where to look for further information

Kansas State University, North Carolina State University, and Auburn University have faculty that focus on feed processing research. Internationally, research is also being conducted at the Chinese Academy of Agricultural Sciences – Feed Research Institute and Wageningen University Animal Nutrition Group. Scientists designing and conducting experiments must consider how feed processing will affect the results of the experiment. Furthermore, feed processing parameters must be included in the materials and methods of all animal nutrition research reports.

11 References

Acosta, J. A., Petry, A. L., Gould, S. A., Jones, C. K., Stark, C. R., Fahrenholz, A. C. and Patience, J. F. 2019. Enhancing digestibility of corn fed to pigs at two stages of growth through management of particle size using a hammermill or roller mill. *Trans. Anim. Sci.* 4(1):10–21 doi:10.1093/tas/txz146.

Amerah, A. M., Ravindran, V., Lentle, R. G. and Thomas, D. G. 2007. Influence of feed particle size and feed form on the performance, energy utilization, digestive tract development, and digesta parameters of broiler starters. *Poult. Sci.* 86(12):2615–2623.

Amerah, A. M., Ravindran, V., Lentle, R. G. and Thomas, D. G. 2008. Influence of feed particle size on the performance, energy utilization, digestive tract development, and digesta parameters of broiler starters fed wheat- and corn-based diets. *Poult. Sci.* 87(11):2320–2328.

ASABE 2012. *Method of Determining and Expressing Fineness of Feed Materials by Sieving. ANSI/ASE S319.4 FEB2008(R2012).* American Society of Agricultural Engineers, St. Joseph, MO.

Chewning, C. G., Stark, C. R. and Brake, J. 2012. Effects of particle size and feed form on broiler performance. *J. Appl. Poult. Res.* 21(4):830–837.

Corzo, A., Mejia, L. and Loar, R. E. II 2011. Effect of pellet quality on various broiler production parameters. *J. Appl. Poult. Res.* 20(1):68–74.

Deaton, J. W., Lott, B. D. and Branton, S. L. 1995. Corn grind size and broilers reared under two temperature conditions. *J. Appl. Poult. Res.* 4(4):402–406.

De Jong, J. A., DeRouchey, J. M., Tokach, M. D., Goodband, R. D., Dritz, S. S., Nelssen, J. L. and McKinney, L. 2012. Effects of feeding varying ingredient particle sizes and diet forms for 25 to 50 Lb nursery pigs on performance, caloric efficiency and economics. Swine Day 2012. Kansas State University Agricultural Experiment Station and Cooperative Extension Service. Progress Report 1038. pp. 305–314.

De Jong, J. A., DeRouchey, J. M., Tokach, M. D., Goodband, R. D., Hastad, C. W. and Dritz, S. S. 2013. Effects of corn particle size, complete diet grinding, and diet form on 24 to 50 lb nursery pig growth performance. Swine Day 2013. Kansas State University Agricultural Experiment Station and Cooperative Extension Service. Progress Report 1092. pp. 95–101.

De Jong, J. A., DeRouchey, J. M., Tokach, M. D., Dritz, S. S., Goodband, R. D., Paulk, C. B., Woodworth, J. C., Jones, C. K. and Stark, C. R. 2016. Effects of wheat source and particle size in meal and pelleted diets on finishing pig growth performance, carcass characteristics, and nutrient digestibility. *J. Anim. Sci.* 94(8):3303–3311.

Engberg, R. M., Hedemann, M. S. and Jensen, B. B. 2002. The influence of grinding and pelleting of feed on the microbial composition and activity in the digestive tract of broiler chickens. *Br. Poult. Sci.* 43(4):569–579.

Evans, C. E., Saensukjaroenphon, M., Sheldon, K. H., Jones, C. K., Paulk, C. B. and Stark, C. R. 2018. The effect of hammermill screen hole diameter and hammer tip speed on particle size and flow ability of ground corn. *J. Anim. Sci.* 96(Suppl. S2):Abstract 319.

Fahrenholz, A. C., McKinney, L. J., Wurth, C. E. and Behnke, K. C. 2010. The importance of defining the method in particle size analysis by sieving. Swine Day Report 2012. Kansas State University Agricultural Experiment Station and Cooperative Extension Service. Progress Report 1038. p. 316.

Fairfield, D. 2005. Pelleting systems and operations. In: Schofield, E. K. (Ed.) *Feed Manufacturing Technology V*, American Feed Industry Association, Arlington, VA. p. 142.

Ferket, P. 2000. Feeding whole grains to poultry improves gut health. *Feedstuffs* 72:12–14.

Gebhardt, J. T., Paulk, C. B., Tokach, M. D., DeRouchey, J. M., Goodband, R. D., Woodworth, J. C., De Jong, J. A., Coble, K. F., Stark, C. R., Jones, C. K. and Dritz, S. S. 2018. Effect of roller mill configuration on growth performance of nursery and finishing pigs and milling characteristics. *J. Anim. Sci.* 96(6):2278–2292.

Goodband, R. D. and Hines, R. H. 1987. The effect of barley particle size on starter and finishing pig performance. *J. Anim. Sci.* 65(Suppl. 1):317–324.

Goodband, R. D., Tokach, R. D. and Nelssen, J. L. 2002. *The Effects of Diet Particle Size on Animal Performance*. Kansas State University Extension Bulletin, MF-2050. Kansas State University Agriculture Experiment Station

Greenwood, M. W., Clark, P. M. and Beyer, R. S. 2004. Effect of fines level on broilers fed two concentrations of dietary lysine from 14 to 30 days of age. *Int. J. Poult. Sci.* 3(7):446–449.

Groesbeck, C. N., Goodband, R. D., Tokach, M. D., Nelssen, J. L., Dritz, S. S., Lawrence, K. R. and Hastad, C. W. 2003. Particle size, mill type, and added fat influence flowability of ground corn. Swine Day Report. Kansas State University Agricultural Experiment Station and Cooperative Extension Service. Progress Report 920. p. 203.

Hanrahan, T. J. 1984. Effect of pellet size and quality on pig performance. *Anim. Feed Sci. Technol.* 10(4):277–283.

Harris, D. D., Tribble, L. F. and Orr, Fr. D. E., Jr. 1979. The effect of meal versus different size pelleted forms of sorghum-soybean meal diets for finishing swine. Proceedings of the 27th Annual Swine Short Course, Texas Tech Univ., Lubbock, TX.

Healy, B. J., Hancock, J. D., Kennedy, G. A., Bramel-Cox, P. J., Behnke, K. C. and Hines, R. H. 1994. Optimum particle size of corn and hard and soft sorghum for nursery pigs. *J. Anim. Sci.* 72(9):2227–2236.

Heiman, M. 2005. Particle size reduction. In: Schofield, E. (Ed.) *Feed Technology V*, American Feed Industry Assn, Arlington, VA. pp. 108–126.

Hu, B., Stark, C. R. and Brake, J. 2012. Evaluation of crumble and pellet quality on broiler performance and gizzard weight. *J. Anim. Vet. Adv.* 11(14):2453–2458.

Idan, F. 2019. The role of feed processing and fiber addition on improving the nutrition and growth performance of broilers. Ph.D. Dissertation, Kansas State University, Manhattan, KS.

Kalivoda, J. R., Jones, C. K. and Stark, C. R. 2017. Impact of varying analytical methodologies on grain particle size determination. *J. Anim. Sci.* 95(1):113–119.

Lemme, A., Wijtten, P. J. A., vanWichen, J., Petri, A. and Langhout, D. J. 2006. Responses of male growing broilers to increasing levels of balanced protein offered as coarse mash or pellets of varying quality. *Poult. Sci.* 85(4):721–730.

Lemons, M. E. and Moritz, J. S. 2016. The effect of feeder space access and crumble- or pellet-to-fine ratio on 38-day-old broiler performance. *J. Appl. Poult. Res.* 25(1):12–20.

Lewis, L. L., Stark, C. R., Fahrenholz, A. C., Bergstrom, J. R. and Jones, C. K. 2015. Evaluation of conditioning time and temperature on gelatinized starch and vitamin retention in a pelleted swine diet. *J. Anim. Sci.* 93(2):615–619.

Martinez, C. M. and Parsons, C. M. 2007. Effect of increased heat processing and particle size on phosphorus bioavailability in corn distillers dried grains with solubles. *Poult. Sci.* 86(2):331–337.

Mavromichalis, I., Hancock, J. D., Senne, B. W., Gugle, T. L., Kennedy, G. A., Hines, R. H. and Wyatt, C. L. 2000. Enzyme supplementation and particle size of wheat in diets for finishing pigs. *J. Anim. Sci.* 78(12):3086–3095.

McKinney, L. J. and Teeter, R. G. 2004. Predicting effective caloric value of nonnutritive factors: I. Pellet quality and II. Prediction of consequential formulation dead zones. *Poult. Sci.* 83(7):1165–1174.

Moritz, J. S., Beyer, R. S., Wilson, K. J., Cramer, K. R., McKinney, L. J. and Fairchild, F. J. 2001. Effect of moisture addition at the mixer to a corn-soybean-based diet on broiler performance. *J. Appl. Poult. Res.* 10(4):347–353.

Nemechek, J. E., Tokach, M. D., Dritz, S. S., Fruge, E. D., Hansen, E. L., Goodband, R. D., DeRouchey, J. M. and Woodworth, J. C. 2015. Effects of diet form and feeder adjustment on growth performance of nursery and finishing pigs. *J Anim Sci* 93(8):4172–4180.

Nir, I., Hillel, R., Ptichi, I. and Shefet, G. 1995. Effect of particle size on performance. 3. Grinding pelleting interactions. *Poult. Sci.* 74(5):771–783.

Nir, I., Twina, Y., Grossman, E. and Nitsan, Z. 1994. Quantitative effects of pelleting on performance, gastrointestinal tract and behaviour of meat-type chickens. *Br. Poult. Sci.* 35(4):589–602.

Ohh, S. J., Allee, G., Behnke, K. C. and Deyoe, C. W. 1983. Effect of particle size of corn and sorghum grain on performance and digestibility of nutrients for weaned pigs. *J. Anim. Sci.* 57(Suppl. 1):260 (Abstr.).

Owens, J. M. and Heimann, M. 1994. Material processing cost center. In: McElhinney, R. R. (Ed.) *Feed Technology IV*, American Feed Industry Assn, Arlington, VA. p. 81.

Pacheco, W. 2014. Effect of particle size and inclusion level of DDGS, and pellet quality on nutrient digestibility, gastrointestinal development, and live performance of broilers and swine. M.S. Thesis, North Carolina State University, Raleigh, NC.

Parsons, A. S., Buchanan, N. P., Blenings, K. P., Wilson, M. E. and Moritz, J. S. 2006. Effect of corn particle size and pellet texture on boiler performance in the growing phase. *J. Appl. Poult. Res.* 15(2):245–255.

Paulk, C. B., Hancock, J. D., Fahrenholz, A. C., Wilson, J. M., McKinney, L. J. and Behnke, K. C. 2015. Effects of sorghum particle size on milling characteristics and growth performance in finishing pigs. *Anim. Feed. Sci. Tech.* 202:75–80.

Pope, J. T. 2018. Non-conditioning factors affecting enzyme thermostability during feed processing. Ph.D. Dissertation, North Carolina State University, Raleigh, NC.

Saensukjaroenphon, M. 2019. The effect of pelleting parameters on phytase stability and pellet quality. Ph.D. Dissertation, Kansas State University.

Schell, T. C. and van Heugten, E. 1998. The effect of pellet quality on growth performance of grower pigs. *J. Anim. Sci.* 76 (Suppl. 1):185.

Skoch, E. R., Behnke, K. C., Deyoe, C. W. and Binder, S. F. 1981. The effect of steam conditioning rate on the pelleting process. *Anim. Feed Sci. Technol.* 6(1):83–90.

Skoch, E. R., Binder, S. F., Deyoe, C. W., Allee, G. L. and Behnke, K. C. 1983. Effects of pelleting conditions on performance of pigs fed a corn-soybean meal diet. *J. Anim. Sci.* 57(4):922–928.

Stark, C. R. 1994. Pellet quality I. Pellet quality and its effects on swine performance. Ph.D. Dissertation, Kansas State University, Manhattan.

Stark, C. R. 2012. Feed processing to maximize feed efficiency. In: Patience, J. F. (Ed) *Feed Efficiency in Swine*, Enfield Publishing, Enfield, NH. pp. 131–149.

Stark, C. R. 2017. *Effect of Particle Size and Feed Form on Broiler Performance*, Asociación Colombiana de Médicos Veterinarios y Zootecnistas Especialistas en Avicultura, Bogota, Columbia.

Stark, C. R. and Chewning, C. G. 2012. The effect of sieve agitators and dispersing agent on the method of determining and expressing fineness of feed materials by sieving. *Anim. Prod. Sci.* 52(1):69–72.

Steidinge, M. U., Goodband, R. D., Tokach, M. D., Dritz, S. S., Nelssen, J. L., McKinney, L. J., Borg, B. S. and Campbell, J. M. 2000. Effects of pelleting and pellet conditioning temperatures on weanling pig performance. *J. Anim. Sci.* 78(12):3014–3018.

Svihus, B., Klovstad, K. H., Perez, V., Zimonja, O., Sahlstrom, S., Schuller, R. B., Jeksrud, W. K. and Prestløkken, E. 2004. Physical and nutritional effects of pelleting of broiler chicken diets made from wheat ground to different coarseness by the use of roller mill and hammer mill. *Anim. Feed Sci. Technol.* 117(3–4):281–293.

Truelock, C. N., Ward, N. E., Wilson, J. W., Stark, C. R. and Paulk, C. B. 2019. Effect of pellet die thickness and conditioning temperature during the pelleting process on phytase stability. Kansas Agricultural Experiment Station Research Reports. Vol. 5(8).

Wondra, K. J., Hancock, J. D., Behnke, K. C., Hines, R. H. and Stark, C. R. 1995. Effect of particle size and pelleting on growth performance, nutrient digestibility, and stomach morphology in finishing pigs. *J. Anim. Sci.* 73(3):757–763.

Xu, Y., Stark, C. R., Ferket, P. R., Williams, C. M., Pacheco, W. J. and Brake, J. 2015a. Effect of dietary coarsely ground corn on broiler live performance, gastrointestinal tract development, apparent ileal digestibility of energy and nitrogen, and digesta particle size distribution and retention time. *Poult. Sci.* 94(1):53–60.

Xu, Y., Stark, C. R., Ferket, P. R., Williams, C. M., Auttawong, S. and Brake, J. 2015b. Effects of dietary coarsely ground corn and litter type on broiler live performance, litter characteristics, gastrointestinal tract development, apparent ileal digestibility of energy and nitrogen, and intestinal morphology. *Poult. Sci.* 94(3):353–361.

Xu, Y., Stark, C. R., Ferket, P. R., Williams, C. M., Nusairat, B. and Brake, J. 2015c. Evaluation of litter type and dietary coarse ground corn inclusion on broiler live performance, gastrointestinal tract development, and litter characteristics. *Poult. Sci.* 94(3):362–370.

Xu, Y., Stark, C. R., Ferket, P. R., Williams, C. M. and Brake, J. 2015d. Effects of feed form and dietary coarse ground corn on broiler live performance, body weight uniformity, relative gizzard weight, excreta nitrogen, and particle size preference behaviors. *Poult. Sci.* 94(7):1549–1556.

Xu, Y., Lin, Y. M., Stark, C. R., Ferket, P. R., Williams, C. M. and Brake, J. 2017. Effects of dietary coarsely ground corn and 3 bedding floor types on broiler live performance, litter characteristics, gizzard and proventriculus weight, and nutrient digestibility. *Poult. Sci.* 96(7):2110–2119.

Chapter 4

Trends in analytical techniques for testing animal feed[1]

F. Debode, J. A. Fernández Pierna, M. C. Lecrenier, P. Veys, O. Fumiere, O. Minet, A. Pissard, J. Hulin, N. Chamberland, and V. Baeten, Walloon Agricultural Research Center (CRA-W), Belgium

1 Introduction

Feed (or feedstuffs) is defined as any substance or product, including additives, whether processed, partially processed or unprocessed, intended to be used for oral feeding to animals (EC Regulation N° 178/2002). Feed includes raw feed materials (e.g. cereals, silage and forage) and compound feeds. The latter are manufactured at the industry or farm level from a mixture of raw materials designed to achieve pre-determined performance objectives among the animals.

Feed production represents a major economic sector. For example, in the European Union, farm animals annually consume around 478 million tons of feed of which 30% is produced by the compound feed manufacturers (approximately 158 million tons). The rest of the feed comes from forages

1 P. Veys wrote the part on microscopy. J. A. Fernández Pierna wrote the part on vibrational spectroscopy with the help of O. Minet for the NIR methods, A. Pissard for the Therahertz and N. Chamberland for portable devices. M. C. Lecrenier presented the proteomic approach and described the operational scheme in link with Figure 2. O. Fumière presented the strategy in link with Figure 1. F. Debode wrote the introduction, the genomic approach, a part of the multidisciplinary approach and the conclusions with the help of J. Hulin for the Data Science section. He made the links between the authors' contributions. All the authors revised the complete manuscript and helped to an easier reading of the text.

http://dx.doi.org/10.19103/AS.2020.0083.10

(approximately 233 million tons) and from feed ingredients (approximately 89 million tons) produced at the farm level (FEFAC, 2016).

Feed analysis is not only used routinely for quality testing as the determination of moisture, protein content, fat content, fatty acids profile, amino acids profile, ash content, digestibility or energetic value of the feed (Baeten et al., 2016), but also to check the safety of materials.

Hazards in feed may present an important risk for human health and can have a negative impact on animal health and welfare (FAO/WHO Expert Meeting on Hazards Associated with Animal Feed, 2015). Risks can be associated with toxic compounds produced by fungal species as aflatoxins and ochratoxins (Streit et al., 2012; Bryden, 2012), bacterial species like *Salmonella* spp., *Escherichia* spp., *Listeria* or *Clostridium* strains (Kadhim and Mater, 2019) or botanical species such as *Datura stramonium*, *Ricinus communis, Ambrosia artemisiifolia* and *Crotalaria* spp. (Aboling et al., 2019; Oseland et al., 2020). Risks can also be associated to forbidden ingredients as processed animal proteins (PAPs) that were banned from the diet of the main farm animals for years to avoid the transmission of spongiform encephalopathies (Horn et al., 2001). Some products, such as GMOs, are associated with specific authorizations and traceability requirements. Fraudulent practices are still ongoing in the feed sector where feed ingredients may be adulterated with other inexpensive and non-authorized materials as it has already been the case with dioxin (Erickson, 1999) or melamine (Jun-Shi, 2009).

The technologies must be adapted to follow a panel of targeted analytes, and so a single technique is not able to answer the different challenges associated with risk detection. Detection methods, based on specific technologies, must also adapt to evolving legislation that can impose new hazards to monitor, new limits of detection or require restrictive authorization. As an example, changes in EU regulations (EC Regulation 56/2013) have led to relaxation of the feed ban and the re-authorization of animal proteins derived from poultry and pigs in feed for aquaculture.

For many decades, the feed chain was presented as just a limited part of the food chain and has probably been neglected regarding its importance and key role in delivering a large quantity of high quality and safe products. New methods were first developed for food control and were only applied to feed in a second step. This was in stark contrast to the risks that have arisen from the feed sector (dioxins, BSE, veterinary residues, etc.). The present and future challenges that the sector faces are the supply of safe, nutritionally optimized, competitive, environmental-friendly and sustainable feed.

The challenges for feed analysis will be to ensure feed quality and safety, to develop and implement innovative strategies to authenticate feed ingredients and feed and to check that they fulfil labeling obligations. This chapter focusses on the authentication of feed and feed ingredients by microscopic, spectrometric, genomic and proteomic techniques.

2 Microscopy

Visual observation of feed ingredients or mixed feed is the very first control which has ever been performed by farmers and feed producers. Such simple visual inspection, in most cases when referred to forage or grains, already delivers a first indication on the quality of a matrix and its composition. It estimates, for instance, the proportions of monocotyledons versus dicotyledons or woody material, type of grains and full-grain mixes, the presence of molds or contamination by foreign bodies such as plastics or metal fragments. However, the limitation of such an approach is rapidly evidenced by the resolution of the naked eye, possibly improved by a magnifying glass, and by the size of the observed fragments. For this reason the use of optical devices such as stereomicroscopes (with magnification ranges up to 50x) was introduced and complemented by microscopes (with usual magnification ranges from 50x to 1000x) for decades, leading to the discipline of feed microscopy. The discipline is supported by reference books such as the 'Gassner', published first in 1931 (Gassner, 1931) and still published in a revised version in 2007 (Hohmann and Gassner, 2007). The republishing of this reference work, after almost 80 years, addresses the question of why this method is still used and simultaneously already gives information on its major limitation, namely the need for expertise.

2.1 Light microscopy

To date, light microscopy is still the most straightforward method for feed control and authenticity. Several reasons account for that. The cost of the equipment is relatively limited, the maintenance of light microscopes is negligible compared to equipment supporting conventional analytical methods and the life duration of the equipment is very long. Furthermore, research or compound microscopes allow a multiplicity of microscopic illumination techniques: bright field, dark field, differential interferential contrast microscopy, phase contrast microscopy, polarized microscopy and fluorescence microscopy. It therefore allows observation of amplitude objects as well as phase objects under transmitted light or in reflection mode offering a wide range of observation conditions for the largest variety of objects. However, making an object visible is only a prerequisite; for control and quality purposes, the accurate identification and categorization of the observed structures are far more essential. This identification based on morphological features of the observed structures, their reactivity to some staining reagents and chemicals, some structural properties such as the birefringence to polarized light and finally the knowledge of the analyst enables the observed structures to be correctly defined. This knowledge can only be acquired by years of experience, by the study of reference micrograph libraries (Belinchón Crespo et al., 2012) or reference materials (most of the time self-collected) and by consultation of reference books (Hahn and Michaelsen, 1996; Klein and Marquard, 2005; Hohmann and Gassner, 2007; Makowski et al., 2010). But this might not be even enough. Effectively the other keystone of microscopic observation is sample preparation, better expressed as the making of slides from a minute amount of a subsample fraction which should reflect the final sample as a whole. This points out the issues of heterogeneity of feed matrices. If one can expect a satisfying homogeneity of a compound feed as results from grinding and mixing the ingredients prior to the extrusion or pelleting process, some other matrices such as hay are almost heterogeneous regarding the size and density of the plant fragments. This is also the case for matrices such as mixtures of grains from different species, which are heterogeneous due to their different species, densities, sizes, shapes and coating of each specific grain type. To overcome this, adequate sampling (Brera et al., 2012) and sample homogenization is required to obtain a representative subsample which will eventually be observed.

Visual observation by stereomicroscopes can be defined as semi-quantitative while microscopic observation, by compound microscopes, deliver results which are qualitative, meaning that they have to be interpreted as binary results (presence or absence of something). Visual observations by stereomicroscopes are made on particles with a size ranging from 250 µm to more than 1 mm, on a large subsample fraction or on the whole sample. At this size, particles can be physically isolated (e.g. by tweezers) and collected for weighing. This will deliver rough estimations (expressed in percentages of weights) but uncertainties of the measurements will still be large, simply because some particles will not be able to be isolated or even identified and hence overlooked. Such situations are classical in the control of declaration for seed mixes, screening of undesirable seeds (Directive 2002/32/EC) or fruits such as achenes of ragweed (*Ambrosia* spp.) (Frick et al., 2011), toxic seeds or seed fragments such as from thorn apple (*Datura* spp.). Any precise quantitative assessment is excluded and therefore such results are classified as semi-quantitative. When approaching microscopic analyses stricto sensu, on particles smaller than 250 µm, the process of isolation of the fragments becomes impossible and only counting of particles based on their recognition by experts can be performed. Although stereology methods (Russ, 2005) offer tools for translating counting into better quantifiable results, these approaches seldom lead to results which are satisfyingly reproducible at least in feed matrices where no automated particle identification is possible and counting strictly relies on human expertise (Veys and Baeten, 2010). In other terms, so far microscopic analyses should be considered as strictly qualitative. An illustration of this is the removal in 2013 (EC Regulation 51/2013) from the European Union legislation of the quantification method (EC Regulation 152/2009) for the detection of processed animal proteins by light microscopy.

Microscopic methods are very sensitive. Detection levels of microscopic methods have shown to be validated through interlaboratory studies at levels as low as 0.0025% w/w for processed animal proteins (Veys et al., 2011). This is far below the required limit of detection inferior to 0.1% for this type of constituent (EC Regulation No 152/2009). To achieve a good sensitivity level, preparation of the sample is usually required to isolate or concentrate the targeted particles. Isolation can be performed according to size by sieving when the targeted particles have regular and different sizes compared to the other untargeted particles from the feed matrix. Isolation by sieving is a traditional practice for larger particles to be observed by stereomicroscopes after their physical selection. But sieving isolation is infrequent when microscopic investigations are required because isolated particles are often too small for individual physical selection and handling. The constraint may be compensated for however by microscopes equipped with laser-pulsed microdissection systems, as recently experienced with feed (Axmann et al., 2015). However, in general, preparation of the sample often then consists of concentrating the targeted fragments by sedimenting the sample by chemicals or mixtures of chemicals at different specific gravities in order to separate fractions according to their range of densities (Makowski et al., 2010; Veys and Baeten, 2018; van Raamsdonk et al., 2019). The different fractions, flotates and sediments obtained are then used to prepare slides. Several successive sedimentations may be performed such as for isolation of insect fragments (Veys and Baeten, 2018) combined with detection of processed animal proteins which are mainly based on the identification of bones having higher specific gravities than the insect cuticle fragments.

However, independently of such sample preparation, the sensitivity of light microscopy for the disclosure of contaminants in feeds or their adulteration with prohibited materials is largely influenced by the operator's skill and expertise. Effectively, the recognition skill of experts is the crucial parameter that will influence both the sensitivity (the ability to correctly detect an analyte effectively present with a low level of false-negative results) and the specificity (the ability to correctly confirm the absence of an analyte effectively absent with a low level of false-positive results). Applied to the disclosure of processed animal proteins, this expertise involves the ability to discriminate fragments originating from different tissues with different histological properties: bones and fish bones, cartilages, hairs, feathers, muscle fibers, gills, teeth, blood, fish scales, echinoderm's stereom, insect cuticles, bivalve shells and others. While the presence of hairs and feathers obviously evidences the presence of mammalian and avian material, in other cases the histological patterns do not allow sorting of the observed particles. For instance, striated muscle fibers from fish, mammals, birds, reptiles, amphibians and, with the noticeable exception of insects (Veys and Baeten, 2018), arthropods cannot be differentiated from each other and are of low value for the control of the species-to-species feed ban. Even bones, the most relevant type of particle used for this purpose, based on the shape, the densities of lacunae and the canaliculi, can only be categorized as 'terrestrial' and 'fish' without any taxonomic significance. Thus, for the detection of animal proteins in feed, although being very sensitive, light microscopy has the limitation of categorization in only two classes, three in the future if the terrestrial category would be split into 'terrestrial vertebrates' and 'terrestrial invertebrates' as proposed (Veys and Baeten, 2018).

Because of the human interpretation, the minute amounts of material being spread on the slides and the impossibility to sort and identify all of the particles on a slide, it is important to observe several slides and possibly repeat observations from a different subsample, as legally imposed (EC Regulation 51/2013), before being able to conclude a result. This arises from the very basic issue of misinterpretation of particles leading to confusion and false-positive results. Furthermore, because of possible intralaboratory carry-over (due to grinding, imperfect cleaning of equipment or air-conditioned rooms) leading to false-positive conclusions, a quantitative threshold of five particles was fixed by the legislator before declaring the result as positive while ensuring that below that threshold the sample would still be below the imposed limit of detection of lower than 0.1% w/w.

Concerning identification by light microscopy of plants composing dry forages, silages or pelleted feed components, it is supported by more reliable and discriminating parameters allowing not only the identification of cereals and common cultivated fodder species but even the identification at the species level of wild meadowland species. Handbooks of references for forages and hay even present identification keys for vegetative plant parts (Rech, 2011) based on the shape, length and surface of plant trichomes of different types - secretory and non-secretory, combined with the shape of stomata guard cells and the number and shape of surrounding epidermal cells, as well as the presence and numbers of calcium oxalate crystals. The larger morphological diversity of plant cells allows a more accurate identification than that possibly achievable for animal structures based on tissues with no or barely differentiated patterns among taxons. This advantage is well documented (Gassner, 1931; Hahn and Michaelsen, 1996; Klein and Marquard, 2005; Hohmann and Gassner, 2007; Makowski et al., 2010) and also applies for the control and identification of other plant parts or products used in feed (cereal and legume grains, cereal meals, starches from various species, roots, fruits and co-products as pulps and many others) or possibly contaminating feed (weed seeds, ground straw and wood dust, noxious and arable species). It must however be mentioned that very little information on the sensitivities and specificities for each plant structure are available, since the recurrent lack of validation studies.

In the future, the use of microscopy will still remain as a first-level method for control and authentication considering the previously cited advantages: cost, life of equipment, ease of use and direct method. However, it needs to be combined with other methods as detailed in other sections of this chapter to obtain further insights on the feed components. This need for a combination is directly linked to the limitations of microscopy: the necessity of human expertise, the impossibility to sort out some materials (mammalian bones, muscles, smooth tissues, products issued from blood for instance) according to their species of origin which is a requirement for the species-to-species feed ban, as well as sometimes the impossibility to visually detect the presence of an ingredient (e.g. plasma powder) once incorporated in a matrix followed by an extrusion process. In 2014, the development of a fluorescence in situ hybridization protocol combining DNA targeting and confirmation by fluorescence microscopy of ruminant bones (Lecrenier et al., 2014) opened new perspectives for species-specific identification of particular components by microscopy.

The simplest way to improve the use of light microscopes, aside from the quality of the optical pathway (quality of objectives), is to optimize micrographic records by a digital camera. High-resolution cameras, over 10 megapixels with pixel resolution of a few μm, are now fairly affordable as well as imaging software. Technological developments in the field of light microscopic image analyses and treatment also reinforce the potential of microscopy. Among the newly available image-enhancing techniques, the stacking of multiple focal plane images into a single extended-focus image, referred to as z-stacking, is a very useful treatment offering better identification of large structures which cannot be focused entirely due to their thickness, and also of small particles for which details can be reconstructed especially when working with 100x immersion objectives which have the shortest working distance, such as for detailed observations of bone lacunae and the network of canaliculi. Based on z-stacking, micrographs showing all structural details should complement official control reports or serve to build up reference collections of micrographs. Image stitching, a technique enabling reconstruction of a single image from multiple fields of observation at high magnification allows high precision views to be obtained from an entire slide in one single file format. Such images obtained, provided they are saved in vector graphic formats, can be exchanged by emailing between experts for discussion at a quality never before obtained.

2.2 Digital microscopy

The recent development of digital microscopy offers new possibilities for feed, feed additives and contaminant detection (Aydin et al., 2020; Ehlers et al.,

2020). With integrated optical systems coupled with image software, the latest generations of digital microscopes allow automated micrographic records and measurements with multiple lighting conditions. 3D reconstructions of surface details enable, for instance, seed fragment analyses without the necessary step of microtomy to achieve adequate thickness before observations by conventional light microscopy. The camera and optical performance allow continuous and progressive observation flows within ranges of magnification from 20x to over 1000x. When powders are analyzed, contamination analysis can be automated based on shape and/or color detection and particle size measurements. Even automatic count functions with export of statistics in conventional spreadsheet formats have become available. This is a series of new microscopic opportunities that is still largely unexplored in feed analysis.

3 Vibrational spectroscopic techniques

Feed safety and quality controls are often performed using reference methods (fundamental chemistry) that have limitations as surveillance and control methods in the feed chain. These methods are (i) time-consuming, while the need of the feed chain is for strategies that are able to give an instantaneous answer; (ii) expensive, while the appropriate safety and quality controls at any of the crucial links of the feed chain require a huge number of analyses to be performed; (iii) realised in the laboratory, while the management control has to be at the production level (online measurement) or directly at the field level (in-field measurement); (iv) inflexible and single purposed (one method/one parameter or analyte), while security and quality control need rapid methods that allow simultaneous analyses of different compounds; (v) sampling-dependent, while the analysis should be representative of the whole product batch; and (vi) not always respectful of the environment (toxic reagents), while the international analytical community looks for minimising the impact of any actions on the environment or health.

Due to the limitations of reference methods in the agricultural sector, there was a real need for the development and validation of fast-screening analytical methods. But there is also a need for non-destructive, smart control systems for quality and safety assessment in a diversity of food and feed industries. This has prompted research teams to develop new analytical solutions, based on spectroscopic technologies (e.g. near-infrared spectroscopy (NIR), mid-infrared spectroscopy (MIR), Raman spectroscopy). The advantages of spectroscopic techniques are speed, ease of use, reasonable start-up cost, non-destructiveness and the possibility of online or in-the-field implementation. Spectroscopic methods enable product control at a much higher frequency, which will upgrade the feed safety and quality control system. Other decisive advantages of spectroscopic methods are the ability to simultaneously determine different factors, without the use of reagents (only required in some extent at the calibration stage) and reduced sample preparation.

3.1 Near-infrared spectroscopy and Raman spectroscopy

Near-Infrared spectroscopy and Raman spectroscopy are techniques that have been applied in a wide range of studies involving food and feed analysis. Some of the main areas of interest involve nutrition, authenticity and traceability of food/feed stuffs in the agricultural sector. These established technologies are coming to the forefront of analysis due to their speed, reproducibility, non-destructive capabilities and in particular due to the relative ease of implementing these low-cost technologies into an industrial setting, that is, as an online or in-line solution. NIR analyses are particularly easy as the preparation of the sample is very limited and most of the time restricted to drying and a grinding steps. Moreover, the costs of NIR analyses by sample are much cheaper, faster and more ecological than other traditional analyses.

NIR has been applied, in the feed and food sectors, to estimate the chemical and physical characteristics of wheat by predicting the nutritive value of grain for broiler chickens (Owens et al., 2007) and to predict the fatty acid composition of beef to include saturated, branched and monounsaturated fatty acids (Sierra et al., 2008). One of the oldest and most successful applications of NIR in the feed area is for the prediction of chemical composition and digestibility in forages, which are the primary self-produced feed resources of ruminants making a significant contribution to the feeding autonomy of the farm (Norris et al., 1976; Sinnaeve et al., 1994; Cozzolino and Labandera, 2002; Boval et al., 2004; Andueza et al., 2011, 2013). NIR is used as a tool allowing determination of the chemical composition and nutritive value of forage production, which is essential not only to adapt the diet to the farm's animal requirements, but also to manage the overall operation of the farm (Park et al., 1998; Tran et al., 2010; Dale et al., 2012; Decruyenaere, 2015). This is where NIR spectroscopy is particularly interesting as it can determine most of the major constituents of forage (ash and organic matter - crude protein, fiber, fat, sugar/starch). NIR has also been applied to control the illegal practice of adulterating animal feedstuff with substandard replacements (Abbas et al., 2013; Qudoos et al., 2019).

In the case of Raman spectroscopy, it has been used to detect oil adulteration in the food industry for quite some time (Baeten et al., 1996; Davis et al., 2000; López-Diez et al., 2003; Zou et al., 2009). However, oil adulteration is not only limited to the food sector but is also likely in the animal feed sector, such that the evaluation of oil used in production of food and feed is become increasingly important, resulting from a series of scares and scandals in the public sector due to contamination with specific compounds (Abbas et al., 2009; Banati, 2011; Furst, 2011). However, only a few studies have used Raman spectroscopy to detect such adulterations (Graham et al., 2012). Raman has been already tested for rapid and routine screening of the presence of melamine in animal feed (Liu et al., 2009) or for the traceability of feed additives (Omar et al., 2017).

The development of robust and flexible spectroscopic instrumentations adapted for online (or in the field) control of the production chain is well-suited for the continuous monitoring of processes from raw materials to finished products. Such systems provide real-time analyses with an increased sample throughput. In this sense, in recent years, new technological developments in NIR spectroscopy include the use of micro electro mechanical system (MEMS) technology. MEMS technology is the integration of mechanical elements, sensors, actuators and electronics on a common silicon substrate through microfabrication technology. This technology brings together silicon-based microelectronics and micromachining technology, allowing the development of smart products, increasing the computational ability of microelectronics with the perception and control capabilities of microsensors and microactuators, and expanding the prospect of possible designs and applications. Many new MEMS applications are emerging; these include spectroscopy in order to meet the increasing demand for online analysis (Fernández Pierna et al., 2010a; Santos et al., 2013).

Miniature handheld instruments are increasingly required in the forage and feed industry (Modroño et al., 2017). Some of them are especially dedicated for forage analysis at farm (e.g. AuroraNIR, NIR4Farm, PoliSPEC). The majority of these handheld NIR spectrometers dedicated to forage analysis have a spectral range between 900 nm and 1700 nm. Application of these devices to forage analysis saves costs (profit margins, accurate ration), time (sample preparation is easier) and they are environmental-friendly. Development of NIR calibrations concerning the chemical constituents of fresh forages is quite challenging comparing to dry forage calibrations (Donnelly et al., 2018). This is mainly due to the high content of water in this material and the high level of heterogeneity (stem, leaves and grains).

Thus, an improvement of these calibrations remains to be done to achieve a level of performance similar to what is obtained with classical NIR. However, handheld NIR spectrometers can give an initial idea of the quality of the forage produced at farm.

3.2 Hyperspectral imaging spectroscopy (HIS) and near-infrared microscopy (NIRM)

Other recent trends in vibrational spectroscopy include the development of methods based on the combination of techniques. In that direction, the different vibrational technologies could be coupled to a microscope creating NIR, MIR or Raman microscopy techniques or to an imaging system to create what is now called hyperspectral imaging spectroscopy (HIS) which allows spectroscopic images to be collected at single kernel or particle levels. This is of great interest for laboratories that control feed compound or cereals.

Near-infrared microscopy (NIRM) is able to collect the spectrum of individual particles (the minimum spot size is a few micrometers) from samples. The instrument includes a classical NIR spectrometer coupled with light microscopy in which the optics has been adapted to NIR radiation. A viewing system magnifies the visible light image of the sample, allowing the user to visualize it and position the infrared beam on the sample area of interest using a motorized stage. NIRM can provide the spectral information of sample at a micro-scale level, so this method can significantly improve the adulterant detecting sensitivity. Moreover, NIRM can provide spatial information of the sample at a micro-scale level, so it can confirm the position of target inside the testing sample for further investigation with other techniques. In the past decade, NIRM imaging has become a powerful analytical tool in quality control and troubleshooting of pharmaceutical products and the detection of contamination and defects in agri-food products. In the feed sector, several works have been published showing the interest of NIRM in the feed sector. In chronological order, NIRM has been applied for the detection and quantification of mammalian PAPs in feedstuffs using raw and sediment fractions (Baeten et al., 2004, 2005) and for the discrimination of fish bones (de la Haba et al., 2007). Abbas et al. proposed some key parameters for a complete development of a NIRM-based method for the quantification of animal by-products (Abbas et al., 2010). In their review, Yang et al. (2011) indicated the potential of near-infrared microscopy to detect, identify and quantify processed animal by-products. Boix et al. (2012) developed a complete validation strategy, through a collaborative study, of the NIRM for the detection of animal products in feedstuffs. The method delivers qualitative results in terms of the presence or absence of animal particles in feed and differentiates animal from vegetable and mineral feed ingredients on the basis of the evaluation of near-infrared spectra obtained from individual particles present in the sample. In the framework of an interlaboratory study, a universal standardization cell for the discrimination of different types of animal proteins has been proposed (Boix et al., 2012; Fernández Pierna et al., 2013). Tena et al. proposed differentiating species using defatted samples (Tena et al., 2014) and differentiation of coccidiostat-containing feed additives was also studied (Omar et al., 2015). Baeten and Dardenne (2016) made a complete review of methods, especially NIRM, to detect contaminants in food and feed. Among those contaminants, special attention has been paid in the literature to the detection and quantification at low levels of melamine in feed (Yang et al., 2014, 2016a; Shen et al., 2020). Also recently, Mandrile et al. (2018) used

NIRM as a potential screening method for the detection and quantification of insect meal in feed by the use of chemometric techniques for the automatic recognition/quantification of insect meal fragments in the feed matrix.

Thanks to the combination of spectroscopy and image analysis, the conformity of products with respect to quality control and the processes of scientific analysis can be constantly optimized. With HIS, it is possible to obtain information not only about the composition of a product, but also about the spatial distribution of its constituents (Gowen et al., 2007). Several books (Baeten et al., 2007; Dale et al., 2013; Fernández Pierna et al., 2020) update the state-of-the-art in the field of hyperspectral imaging.

HIS has been largely used in the last years in the agri-food sector. In the feed sector, several studies have been performed. This technique has been applied for safety and reliable control of feed components. In this direction, Wang and Paliwal (2005) proposed the use of a hyperspectral imaging system working in the near-infrared wavelength region combined with chemometrics to study the possibility of detection of ground-broiler feed contamination by ground pork (Wang and Paliwal, 2005). HIS was also included in a complete framework for developing and validating a method as a standard protocol accepted by regulatory authorities for the detection of PAPs in feedstuffs. In that work, studies aimed at determining the performance characteristics and robustness including limit of detection, repeatability and the risk of cross-contamination were included (Fernández Pierna et al., 2010b).

One potential area of application for HIS is the detection of extraneous material and hazards in food or feed products. HIS is especially suited for grains or solid granular material, as once a proper classification model is built, automatic recognition of different objects can be done by attributing each pixel to a particular class (Pearson and Wicklow, 2006; Williams et al., 2012; Tähkäpää et al., 2015; Chu et al., 2017; Kimuli et al., 2018a,b; Tao et al., 2018). Caporaso et al. (2018) made a complete review of applications of HIS for quality assessment of cereal grains. In the same context, studies have been done in safety control of plant-based food that is converted to by-product and used as animal feed (Colovic et al., 2014).

HIS techniques have been used for the detection of animal proteins in compound feed in order to compile with legislation regarding the use of processed animal proteins in animal nutrition. In this area, these methods have also been used for developing methods for the full screening of all kind of particles in ground compound feed (Baeten and Dardenne, 2016) or for the quantification of protein content (Morales Sillero et al., 2018) among others. They have also been proposed for detecting other types of plant and mineral contaminants. HIS has also been used successfully to detect, at the ppm level, the presence of ergot bodies and other plant contaminants, or botanical impurities in cereals. One study has shown that HIS technology can be used to detect and quantify ergot contamination in cereals intended for food or feed, in both the laboratory and the production chain, using online prediction (Vermeulen et al., 2012; Vermeulen et al., 2017), showing that HIS could be a powerful tool within the framework of official controls (Vermeulen et al., 2013).

3.3 Chemometrics

The use of vibrational spectroscopic techniques has been driven in parallel with the huge developments that have been achieved in chemometrics. Chemometrics is the chemical discipline that uses mathematics and statistics to design or select optimal experimental procedures, to provide maximum relevant chemical information by analyzing chemical data, and to obtain knowledge about chemical systems (Massart et al., 1997). Main improvements have been achieved in (i) data reduction allowing the acceleration of data analysis, (ii) construction and management of data bases, (iii) the combination of data from different techniques, (iv) development of algorithms to perform appropriate extraction and exploitation of the data and (v) interpretation and presentation of the results (Martens and Naes, 1989). These features greatly benefit from the incredible development of computer capabilities and internet facilities.

Compliance to international standards, especially for method validation and coherence of the numerous measurements along the traceability systems, is mandatory (Baeten et al., 2010). Today it is not logical to develop methods that do not meet the requirements of international standards for use in a quality system environment. In this direction, significant effort has been made to adapt the use of spectroscopic techniques and chemometric developments for the correct application of such standards and legal rules in the area of safety and quality of the feed chain.

3.4 Terahertz (THz)

As a trend in the spectroscopic domain, new powerful techniques are emerging. Techniques such as Terahertz (THz) applied to complex molecules forming food/feed matrices have been investigated in connection with several issues. THz spectroscopy exploits the interaction of THz radiation with matter in the gaseous, liquid or solid states (Theuer et al., 2011). THz has been explored for adulteration issues such as the detection of melamine (illegally added to feed and food products to enhance their apparent protein content) (Ung et al., 2009; Baek et al., 2014) or adulteration of oils (Zhan et al., 2016). Contamination issues have also been investigated such as the detection of insects in grains (Tan et al., 2014). Moreover, THz has been used to study water-linked issues. Moisture is related to many properties that affect the quality of feed and food but also its shelf life. Some authors succeeded in calibrating THz response with moisture content (Parasoglou et al., 2009). Finally, composition issues could also be addressed by THz spectroscopy. In the case of feed composition, Qin et al. investigated THz spectroscopy as a rapid and sensitive detection technique of residues of an antibiotic used as feed additive (Qin et al., 2015). However, most studies concerning isolation of molecules are not considering the matrix effect and then the risk of change or loss of crystallized forms, which are frequently responsible for the generation of spectral THz features in solid samples. Moreover, studies are mainly based on optical constants and no extensive work focuses on spectral interpretation and the determination of characteristic peaks - in other words, on proper THz characterization. Databases need to be developed to describe the molecular signature of different feed compounds, and band attribution is required in order to use spectra correctly. It is obvious, at this time of publication, that there are great challenges and a lot of work still needs to be done with this spectroscopic technique.

4 Genomic methods

Genomic methods allow the detection of ingredients or contaminants containing genetic information. The genomic tests target a specific nucleic acid sequence of the organism, either DNA or RNA. The majority of the techniques is based on amplification methods as it allows the detection of low amounts of the target and can be used for routine analysis.

4.1 Polymerase chain reaction (PCR)

Most of the detection methods include the polymerase chain reaction (PCR) in their methodology (Mullis, 1990; Levin et al., 2018; Lo and Shaw, 2018). PCR consists of the enzymatic multiplication of a targeted DNA fragment. Genomic methods will then require a DNA extraction step. DNA is a robust molecule that can be recovered from processed feed even if the technical process (grinding, heating, use of solvents for oil extraction) will impact the mean size of the DNA fragments (Fumière et al., 2006; Debode et al., 2007).

Different DNA extraction methods exist (the most frequently used being solvent-based methods, DNA-binding silica columns and silica-coated magnetic beads) and will differ in terms of cost, time, DNA yields obtained and purity (Demeke and Jenkins, 2010; Barbosa et al., 2016). DNA extraction from feed is generally not a problem (Piskata et al., 2019). Simplified DNA extraction methods also exist but the results can be impacted due to the presence of remaining inhibitors that will affect the enzymatic reaction of amplification.

PCR allows the detection of a specific DNA fragment, thanks to its exponential amplification. This makes the technology very sensitive but imposes precautions to avoid contaminations, since from a single molecule billions of copies can be produced. Knowledge of the sequence to amplify is necessary to design complementary oligonucleotides (primers) located at the extremities of the fragment to amplify. The primers will have two functions: identifying by hybridization the region of the genome to amplify and setting the starting point for DNA synthesis to the DNA polymerase which can only add new nucleotides to an already existing strand of DNA.

The techniques will take advantage of the properties of DNA that is able to dehybrid and rehybrid depending on the temperature. The PCR reaction occurs in a thermocycler, a machine performing thermal cycles, each cycle containing 3 steps. At the step of denaturation (92–95°C), the two strands of DNA will separate, at the hybridation step (55–65°C depending of the length and sequence of the primers), the primers will hybridize and at the elongation step (around 72°C), the DNA polymerase will synthesize a new strand from a hybridized primer. The product of amplification is visualized after migration on an agarose gel in the presence of an intercalant reagent (generally ethidium bromide or Sybrsafe). Classical agarose gels are being replaced by capillary electrophoresis. This improvement allows this technique to be kept in the interest of traceability and for authentication purposes (Acunha et al., 2016; Lian and Zeng, 2017; Álvarez-Rivera et al., 2018). The power of PCR was largely exploited for the detection of a large panel of organisms in feed products. The problem of 'classical' PCR is that it is only possible to provide a qualitative result and say if the targeted organism is present or absent. That is why classical PCR was replaced with the emergence of the real-time PCR.

4.2 Real-time polymerase chain reaction (PCR)

Real-time PCR, also called quantitative PCR (qPCR) allows the amplification kinetic to be followed, thanks to a fluorescence emission which is proportional to the quantity of amplicons produced (Higuchi et al., 1993). The real-time PCR machine is a thermal cycler coupled to a camera that measures the fluorescent signal and to a computer that records the progress of the amplification and analyzes the results. The fluorescence can be provided by using an intercalating reagent such as SYBR® Green, and the PCR test can then be easily transferred from classical PCR to real-time PCR. However, since the fluorescent molecules can also intercalate between primer-dimers, some late non-specific signals can be observed but are distinguishable from positive amplification signals with a dissociation protocol checking the melting temperature (Tm) of the amplicons at the end of the run.

For this reason the format using hybridization probes to follow the fluorescence emitted is generally preferred. Moreover, a better specificity is generally obtained by the introduction of the third oligonuclotide that will hybridize between the primers. Double dye probes (Holland et al., 1991) are the most used. The probe is labelled by two dyes and remains non-fluorescent when not degraded. During the elongation step, the polymerase will degrade the probe on the way, thanks to its exonuclease activity, the dyes will be released and the fluorescence measured.

Since 2000, detection methods by real-time PCR were proposed in food and feed (Lockley and Bardsley, 2000) and the development is still ongoing for numerous and various organisms such as plants (Debode et al., 2018; Campos et al., 2018), including the genetically modified ones (Debode et al., 2013; Debode et al., 2018b), animal species (Golinelli et al., 2016; Debode et al, 2017), viruses (Dee et al., 2018), bacteria producing vitamin B2 (Paracchini et al., 2017), probiotics (Sattler et al., 2014; Fibi et al., 2016), pathogenic bacteria (Bassi et al., 2013; Liu et al., 2017) or fungi producing mycotoxins (Scauflaire et al., 2012; Okorski et al., 2017). Short targets below 120 bp are preferred to provide an efficient amplification by real-time PCR in processed feed (Fumière et al., 2006; Debode et al., 2017b).

Real-time PCR is considered as the gold standard method for the detection of specific sequences in a sample. Numerous performance criteria have been described to validate or implement methods in the laboratory (Broeders et al., 2014; ENGL, 2015, 2017; Debode et al., 2017c; Shehata et al., 2019) and method developers have to show that their method is able to meet minimum performance criteria for efficiency of amplification, specificity (detection of the organisms targeted and absence of cross-reactivity with other organisms), sensitivity (ability to detect a level of 0.1% in mixed samples and at least 20 copies of the target) and robustness (detection of 20 copies of the target even if slight modifications are brought to the protocol of detection).

Multiplexing (amplification of several targets in a same tube) is possible by combining different couples of primers and probes and dyes emitting at different wavelengths. Amplification of the different targets can be visualized by selecting the appropriate wavelength. No more than five targets can be amplified in a single tube if correct separation of the signals is expected (Garrido et al., 2012, 2013; Huber et al., 2013; Kim and Kim, 2017).

Portable real-time PCR machines have been developed (Fernández-Carballo et al., 2016; Furutani et al., 2017). However, fast DNA extraction in the field remains less efficient than in the laboratory due to a higher content of inhibitors in the DNA extract. Simplified machines that do not allow modifications of the thermal program with dedicated positions for the controls and unknown samples have also been developed for quality control in the food and feed industries. These machines are less precise but allow a rapid decision before releasing a batch.

PCR remains a targeted approach while untargeted approaches are more and more sought. Real-time PCR allows quantitative or semi-quantitative approaches but requires calibration curves built with certified reference materials at known concentrations or copy numbers. Digital PCR allows a more direct quantitation.

4.3 Digital polymerase chain reaction (PCR)

Digital PCR is based on partitioning of the reaction mix into a high number of small volumes. The division can be carried out by distributing the reaction mixture in hundreds or thousands of chambers of a nanofluidic plate (digital chamber PCR or dcPCR) or in a very high number of droplets within an emulsion (digital droplet PCR or ddPCR). The partitions are so small that with an adequate dilution of the DNA sample, they will contain between one and three copies of the target. A PCR will occur in each partition and will result in the presence/absence (reflecting the name 'digital', 1 or 0) of an amplification reaction. By counting the partitions giving positive and negative results, the number of copies present in the DNA extract used may be quantified. As several copies could sometimes be present in the same partition, Poisson statistics are applied to correct the values. This technique is of great interest because it is less susceptible to the presence of inhibitors and able to detect and quantify low numbers of copies more precisely. Many detection methods are proposed for GMO (Morisset et al., 2013; Košir et al., 2019;), animal species (Cai et al., 2017) or bacteria (Yang et al., 2016b; Jang et al., 2017; Pacocha et al., 2019; Zhao et al., 2017; Selvaraj et al., 2019) in food and feed.

As for real-time PCR, development of multiplex approaches (Dobnik et al., 2016; Košir et al., 2017; Köppel et al., 2019) and portable solutions (Gou et al., 2018) is ongoing.

4.4 Isothermal amplification

Isothermal amplification is an alternative to classical PCR. The technology does not require costly and heavy devices nor thermal cycles for amplification and is easy to use. However isothermal amplification only allows qualitative results and developments are more complicated than for PCR. Several variations exist, the most used being LAMP (*loop-mediated isothermal amplification*) for which many applications have been proposed in food and feed (Nagamine et al., 2002; Song et al., 2017; Li et al., 2018; Niessen, 2018; Domesle et al., 2018; Liu et al., 2018; Yang et al., 2018; Zhang et al., 2018; Long et al., 2019; Shan et al., 2019; Kumari et al., 2019). LAMP amplifies DNA at constant isothermal conditions varying from 60°C to 65°C with specially designed primers in the presence of an intercaling reagent such as SYBR I that will generate a fluorescence in the tube when the target is present. LAMP is interesting for fast on-site detections.

4.5 Sequencing-by-synthesis

4.5.1 Sanger sequencing and pyrosequencing

Delivering the sequence of a targeted region provides the proof of the presence of the organism. Classical Sanger sequencing (Sanger et al., 1977) or sequencing-by-synthesis such as pyrosequencing (Nyrén et al., 1993; Fakruddin and Chowdhury, 2012; Shang et al., 2017) can rapidly provide a sequence. However, they are little used in feed analysis because they are not adapted to the presence of mixed products or microorganisms having similar sequences. Sequencing can be used for the detection of a specific sequence (Jonkuviene et al., 2013; De Battisti et al., 2014; Debode et al., 2017d; Shenge et al., 2019) but increasingly supports high throughput sequencing when reference sequences for DNA barcodes are needed, as explained hereunder.

4.5.2 High throughput sequencing (HTS)

High throughput sequencing (HTS) allows the sequencing of several million fragments at the same time. It already allowed the complete sequence of several organisms to be obtained (Michael and Jackson, 2013; Dangel et al., 2019). However, problems of coverage are encountered with whole-genome sequencing approaches when applied to organisms having a large genome or on mixed products (Willems et al., 2016).

That is why sequencing approaches were developed to focus on sequences of interest. The closest to be used as routine analyses were proposed for GMO and plant species detection. Enrichment of sequences of interest was realized by sequencing a large number of amplicons (Arulandhu et al., 2018) or the use of capture probes to catch targeted fragments to be sent for sequencing (Debode et al., 2019). These approaches remain targeted but hundreds of targets can be tested in one analysis. In the field of GMO detection, the ability to detect percentage as low as 0.1% has already been demonstrated (Debode et al., 2019).

Very active research has also been launched on metabarcoding approaches that will in the future allow detection of all the ingredients within a compound feed product in one single analysis. Targets can be oriented against bacteria, fungi, plants and animals (Mayo et al., 2014; Littlefair and Clare, 2016; Staats et al., 2016; Carvalho et al., 2017) and could all be performed in one single analysis. They use barcodes (600–300 bp) or minibarcodes (<300 bp). Barcodes are amplified with identical primers but the sequence internal to the primers allows for a distinction between organisms. Due to the huge variety of organisms, some markers will not give the desired resolution and do not allow to identify organisms to species (Speranskaya et al., 2018).

Semi-quantification will remain difficult with HTS as the different ingredients of a feed product can result from different processing stages and will have a large variability in terms of DNA quality and availability.

The price and expertise remain relatively high at the moment for feed analysis (EFSA, 2018). Price per sample can be decreased as the DNA of each sample can be tagged with specific sequences prior to mixing and the sequences demultiplexed after sequencing and attributed to the right sample. Assignment of the sequences to the correct organisms is still a challenge as the reference database is still to be constructed and the diversity of organisms can be very high. Bioinformatics treatments and the construction of databases are very important points (Barbosa et al., 2017; Haynes et al., 2019) and are the subject of a lot of attention from the developers due to the high potential of the technology. Portable devices, of the size of a USB key and based on the nanopore sequencing technology, have been developed and their performance is increasing over the years (Fraiture et al., 2018). This cheaper technology should attract new applications for food and feed characterization in the coming years.

5 Proteomics

Proteomics is defined as the study of the proteins of an organism, the proteome, just as genomics studies its genome. The proteome reflects gene expression and the post-translational modifications (PTMs). Both are influenced by internal conditions such as cell type, the stage in the life cycle and its related physiology or by external conditions such as environmental conditions, feeding, handling, harvesting or slaughtering.

In the particular case of feedstuffs, technical processing of materials will have a huge impact on the feed proteome. Various treatments such as cooking, drying, rendering, extrusion or silage will affect the protein content by modification, denaturation, degradation or destruction (Raes et al., 2018). Among others, oxidation, carbonylation, hydroxylation, Maillard glycation, deamidation are frequent PTMs (Ortea et al., 2016). These considerations can make development of the method harder but this characteristic can also be a real advantage in the context of feed authentication as it can be used to draw a specific profile, not only of the initial phenotype, but also the real state of the feed materials (Clerens et al., 2012).

5.1 Immunoassays

The two main techniques used in proteomics are based on immunoassays or mass spectrometry. As reviewed by Gomez-Morte et al. (2019), immunoassays, and the enzyme-linked immunosorbent assay (ELISA) in particular, are commonly used in food sciences due to their high specificity based on the detection of the targeted epitope. ELISA is a highly sensitive technique, with, for example, a detection limit for food allergens ranging from 0.1 ppm to 5 ppm (Croote and Quake, 2016). Other advantages of the technique are that they are mostly cheap, easy to use and fast. However, its main drawback is its loss of sensitivity in case of modifications of the conformational structure, due to protein denaturation, making this technique less suited to processed food (Montowska et al., 2014).

5.2 Mass spectrometry (MS)

In mass spectrometry (MS), protein identification is based on the mass to charge ratio (m/z) of its primary structure, the amino acid sequence. MS reaches a similar level of sensitivity as immunoassays in unprocessed or low processed ingredients but has the advantage of being less sensitive to the loss of protein conformation (Dumont et al., 2010; Planque et al., 2017), making the MS approach a method of choice for processed samples.

This section focused on the mass spectrometry-based approach. MS analysis can be summarized in three steps: (1) ionization of the molecules (proteins or peptides) (2) separation and/or selection of the ions based on their m/z by mass analyzers and (3) ion detection and recording of the ion number of each m/z.

Proteins are polar, non-volatile and thermally unstable molecules requiring soft ionization techniques in order to convert them into a gaseous state while preserving them from significant degradation (Yates et al., 2009). Matrix-assisted laser desorption ionization (MALDI) and electrospray ionization (ESI) are the techniques commonly used in proteomics. The most common mass analyzers associated are (1) quadrupole mass analyzers (Q), which use m/z stability to separate the ions, (2) ion traps, which separate ions based on their m/z resonance frequency, and (3) time of flight (TOF) analyzers, which use the ions' flight time. Each mass analyzer type has unique properties, such as resolution, mass accuracy and sensitivity.

The principle of tandem mass spectrometry (MS/MS) is the combination of two analyzers separated by a collision cell in a single instrument. In proteomics, fragmentation of the proteins preferentially affects the peptides' bonds so that the analysis of the fragments gives information about the amino acid sequence of the peptide. Depending on type of mass analyzers combined, the tandem mass spectrometer will have different features in term of resolution, mass accuracy and sensitivity and, in this way, different applications. A triple quadrupole, having low resolution and mass accuracy but high sensitivity, will be particularly well suited for use in routine analysis by allowing rapid and continuous monitoring of the specific peptides of interest. On the other side, QTOF will be used for method development since the instrument will benefit from the high resolution and mass accuracy of the TOF, allowing high-resolution mass spectrometry (HRMS) to be performed. But this system will have middle sensitivity. The use of MS based-proteomics approach has drastically increased in recent years, due to the significant advances in MS technologies, both in terms of ionization techniques and mass analyzers (Korte and Brockmeyer, 2017), with unique properties which have to be combined depending on the selected strategy and information needed (Cunsolo et al., 2014).

MS-based proteomics has already demonstrated its usefulness in food analysis in different topics such as quality, authentication and safety purposes (Sentandreu and Sentandreu, 2011; Clerens et al., 2012; D'Alessandro and Zolla, 2012; Cunsolo et al., 2014; Ortea et al., 2016; Piras et al., 2016; Colgrave, 2017). But when we review publications of proteomics applied to animal feed, it is clear that this use is still at its very beginning.

Most of the publications propose its use to increase the quality of animal production (e.g. milk, meat), or to measure the impact of animal welfare and health on animal production (Bassols et al., 2014; Marco-Ramell et al., 2016; Kusebauch et al., 2018).

In recent decades, the food chain has faced several major crises, caused by contaminations in the feed part of the chain. The Bovine spongiform encephalopathy outbreak, better known as 'Mad cow disease', is one of them. The reason for this epidemic was the use in cattle feed of ruminant meat and bone meals produced from sick animals following a process deviating from the prescribed conditions of temperature and pressure. Therefore, the use of such animal by-products was strictly regulated. It is therefore not surprising to find the major development of proteomics in feed for the detection of animal by-products. As it is applied in meat authentication (Montowska and Pospiech, 2013; Von Bargen et al., 2013), proteomics is particularly suitable for identifying prohibited animal proteins in feedstuffs because it provides information both about the tissue and species of origin. Ocaña et al. (2004) have first worked on the topic for gelatine detection by MALDI-TOF MS and liquid chromatography-electrospray ionization-tandem mass spectrometry (LC-ESI-MS/MS). Buckley et al. (2009) have then also used MALDI-TOF MS for the specific identification of bone collagen. With the view of a progressive lifting of the feed ban, the development of alternative analytical methods is increasing again and proteomics was one of the those investigated (Reece et al., 2012). The general approach was relatively similar for the different research groups by using LC-ESI-MS/MS, proteome profiling using high-resolution instruments for the identification of peptide biomarkers and subsequent targeted multiple reaction monitoring (MRM) design (Lecrenier et al., 2016; Marbaix et al., 2016; Marchis et al., 2017; Lecrenier et al., 2018; Niedzwiecka et al., 2018; Steinhilber et al., 2018, 2019).

In the same way, proteomics can also be used for authentication of feed materials of plant origin. Feed protein content is often considered as a good indicator of quality and is based on crude protein calculation using the Kjeldahl method. But this estimation does not give information about the protein type and, therefore, the source of protein is not evaluated. By analyzing the protein profile, proteomics can detect a modification in the feed ingredients used. Moreover, conservation conditions, in particular ensiling, have a high impact on the forage proteome (Dong et al., 2019). During ensiling, for instance, proteins are digested by microbial protease, which represents a loss of nutritional value. This degradation varies depending on the botanical composition (Hart et al., 2016) and the ensiling conditions. Beside forage, soybean meal also plays an important role in animal feed. Seo and Cho (2016) have identified the potential use of proteomics tools to follow changes in the protein profile, in particular allergenic and anti-nutritional proteins, during soybean fermentation. Maria John et al. (2017) has also used proteomics to study the influence of genotype, environment (e.g. planting, sowing season) and the relationship between both on anti-nutritional factors. This demonstrates that proteomics is a major method to be used to evaluate feed quality and for its authentication.

All these studies were possible, thanks to pre-existing knowledge of such products and the availability of the proteome database. But, for some others, the protein sequence database is highly incomplete. One current example is the insect protein database. The feed sector is in constant evolution to search for novel sources of proteins. Insects have been identified as an interesting source of proteins for animal feed. Therefore, analytical approaches are required to monitor fraudulent use of insects in feedstuffs. But the lack of proteome data from many insects do not permit the classical proteomics approach based on the identification of biomarkers. Therefore, untargeted approaches are in development. This consists of the construction of spectral libraries by HRMS to differentiate insect species by unsupervised statistical analysis (Belghit et al., 2019; Francis et al., 2020). The creation of an extensive spectral library collection could be a solution to continue the development of proteomics in feed (Rasinger et al., 2016).

A common trend in analytical methods is the development of more sustainable approaches by looking to 'faster' and 'greener' analyses. In the context of MS analysis, the current revolution is linked to sample preparation since it is often highly time-consuming and chemical-consuming. Classically, two workflows are possible in proteomics: bottom-up and top-down. Sample preparation will depend on the approach chosen. Bottom-up is the most applied and sample preparation consists of four critical steps: protein extraction, protein denaturation, enzymatic digestion and clean-up of peptides. In the top-down strategy, proteins are not digested. The solution will probably come from a combination of sampling and ionization by using ambient ionization techniques (Ouyang and Cooks, 2009), requiring minimal sample preparation. For the last 10 years, development of this technique has skyrocketed and, according to Lu et al. (2018), more than 40 kinds of ambient ionization methods have already been developed and used in food science and industry. In their review, Feider et al. (2019) have listed those as most used and categorized them according to their desorption method. The three major types are liquid extraction, plasma desorption and laser ablation. While ambient ionization MS was already broadly used in metabolomics and lipidomics, it is less the case for the proteomics application. Here again, the two workflows are possible with, in case of the bottom-up approach, an initial digestion step before ambient ionization. In the context of food analysis, liquid extraction surface analysis mass spectrometry (LESA-MS) was developed for the identification of meat species (Montowska et al., 2014, 2015). Feider et al. (2019) also give some examples of ambient ionization MS imaging, including a combination of light microscopy and nano desorption electrospray ionization (Nano-DESI) to perform guided protein profiling in tissue (Hsu et al., 2017). As described in a previous section, light microscopy is still widely used in feed control, and the combination with proteomics tools would allow direct protein profiling at a microscopic scale.

6 Multidisciplinary approach

In the past, developments of methods were generally proposed by people who were specialized in an analytical domain and focused on a single technology. A multidisciplinary approach was possible, thanks to the arrival of machines combining two different technologies (i.e. hyphenated techniques). An example is the combination of light microscopy and NIR that gave birth to NIR microscopy where a spectrum was obtained for each particle targeted by the microscope (Baeten et al., 2005; Abbas et al., 2010; Yang et al., 2016a). The technology later evolved to the NIR hyperspectral camera where a spectrum was obtained for each pixel of the scene analyzed (Fernández Pierna et al., 2012; Vermeulen et al., 2013). Due to the huge number of infrared spectra produced with such technologies and the spectral database coming from thousands of samples, statistical and chemometric treatments were mandatory (Fernández Pierna et al., 2012). This also allows the use of a first screening approach using an untargeted class model, followed by a target-specific regression model allowing, for example, to assess the amount of each specific adulterant with better sensitivity (Capuano et al., 2015). The increased application of spectroscopy techniques is closely linked to the development of chemometrics. A similar trend was observed in molecular biology where new approaches by high throughput sequencing or mass spectrometry required the acquisition of bioinformatics skills or the help of bioinformaticians. Automation of analysis is also actively searched for such as in light microscopy combined with image analysis (Pinotti et al., 2015; Ottoboni and Pinotti, 2017).

Besides the multidisciplinary approach required for data management, a combination of technologies and skills also occurs for confirmatory approaches, faster analysis or finding solutions for complex issues. A first example is the association of the classical techniques of microbiology for growing and separation of bacteria on selective media, rapid characterisation of the strains by MALDI-TOF mass spectrometry on a Maldi Biotyper (Bruker Daltonics, Billerica, MA) and confirmation by DNA sequencing of a 16S fragment (Bultreys and Gheysen, 2020). The aim of the MALDI-TOF is to have fast identification directly on bacterial colonies within 15 min. The rapid identification is possible by the comparison of the obtained spectrum to the spectra database.

The evolution of an analytical scheme for the detection of PAPs in feedstuffs gives other examples of strategies that combine several techniques (Lecrenier, 2019). The composition of feed can be very variable to answer to the nutritional needs of each species at the different stages of their life. Nowadays, the list of ingredients is continuously increasing, especially to diversify the sources of proteins. By combining the advantages of each method in an operational scheme to fulfil possible analytical gaps, a more complete answer can be given. In the past, control of the feed ban was based on light microscopy exclusively. It consisted of identification of specific particle types. Indeed, whatever the origin of the tissues, if any animal particle was detected in a feed for farm animals, the feed was considered, with minor exceptions, as containing unauthorized by-products. In 2013, thanks to the decrease of bovine spongiform encephalopathy (BSE) cases in Europe, a partial relaxation was possible with the reintroduction of non-ruminant PAPs in aquafeed. The notion of 'species of origin' became the main challenge. The same year, PCR was officially added to light microscopy as a reference method for the detection of ruminant DNA in PAPs and the current analytical scheme combining light microscopy and PCR was proposed (EC Regulation 51/2013). A few years ago, it was already suspected that authorized ingredients of animal origin would be a limitation. Different strategies were attempted to by-pass the problem. One of them consisted of replacing light microscopy with near-infrared microscopy to increase and automate the number of particles observed in combination with PCR analysis of animal particles taken separately. The first challenge was the recovery of the suspected animal particles. Axmann et al. (2015) investigated the use of a microscope coupled with a laser microdissection and pressure catapulting system (LMPC) for the tedious step of isolation of the particles without risk of contamination. The second challenge was the DNA extraction from such a small amount of material that comprises a single particle. Dedicated

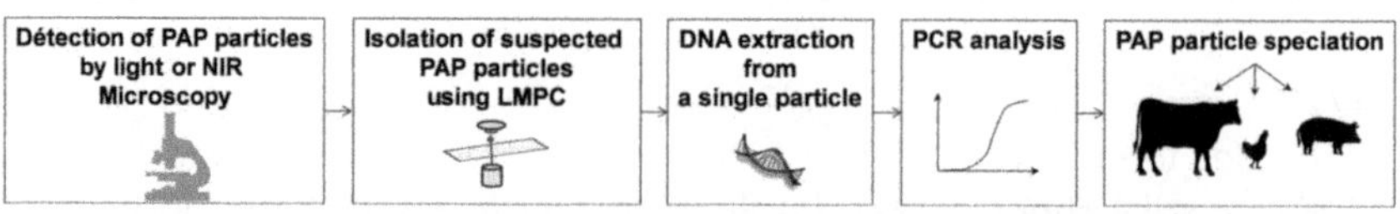

Figure 1 Operational scheme for PAP particle speciation combining light or NIR microscopy with microdissection and PCR. (Legend: PAP = Processed Animal Protein, NIR = Near Infrared, LMPC = Laser Microdissection and Pressure Catapulting, DNA = DeoxyriboNucleic Acid, PCR = Polymerase Chain reaction).

protocols often using forensic techniques were adapted and tested (Fumière et al., 2008, 2010). The workflow is presented in Fig. 1.

Lecrenier et al. (2014) developed another attractive approach by using light microscopy coupled with a fluorescence in situ hybridization (FISH) protocol to determine the ruminant origin of bone particles.

But, since 2013, some positive results for ruminant DNA have been reported in aquafeed. In some cases, the possibility of 'false-positive results' due to the use or presence of dairy products, an authorized ingredient, has been advanced. This has underlined the necessity to provide species and tissue-specific results. A solution to bridge the current analytical gap (presented by the cross in Fig. 2) is the use of mass spectrometry-based proteomics for the detection of unauthorized proteins of ruminant origin (Lecrenier et al., 2020). This third method is currently under evaluation and will probably come to complement the analytical scheme in the future.

These examples show that a multidisciplinary approach can be beneficial to find solutions. The cost and time required must also be taken into account as the price of multiple analyses must remain affordable and realistic. This is also a change in mentalities as experts often promote their technology against other existing methodologies while most appropriate solutions are more from a collaborative approach integrating the strengths and weakness of each technology. Nowadays, more laboratories host several technologies that can be used to analyse the same sample. Merging data of the results coming from different techniques can be a powerful tool for obtaining more reliable results for description, classification or prediction of the samples and will probably gain in importance in the future (Borras et al., 2015).

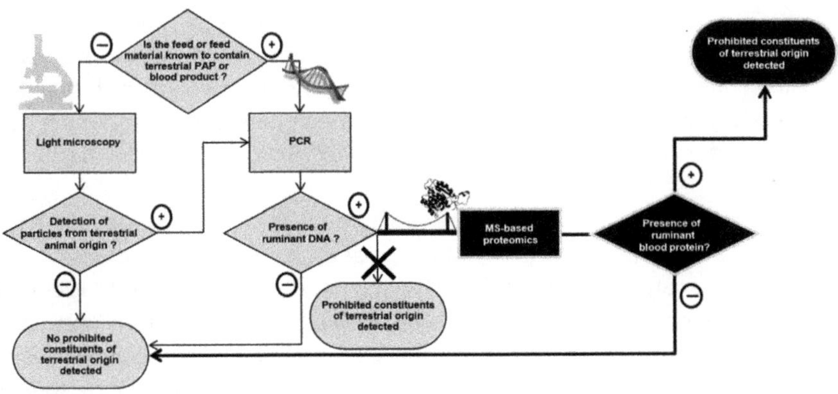

Figure 2 Operational scheme for PAP authentication including light microscopy, PCR and MS-based proteomics (Legend: PAP = Processed Animal Protein, DNA = DeoxyriboNucleic Acid, PCR = Polymerase Chain reaction, MS = Mass Spectrometry).

7 Conclusion and future trends

This chapter focused on light microscopy, vibrational spectroscopy, genomic and proteomic approaches. All these technologies are rapidly evolving. This is also the case for the chromatographic, isotopic, elemental and immunological techniques that were not included in this chapter. New technological developments are generally first proposed for spatial research, medical research and chemical research. After some years they are integrated into food control and finally in feed. The period before implementation of an innovative technology in the feed sector is however decreasing year after year.

The improvement of technologies has different purposes. There is an increasing interest for fast technologies in order to reduce to the minimum the time lapse between acquisition of the information and the decision in order to act rapidly on the process. Spectroscopy has a major advantage on this point. This technology is non-destructive, can be applied directly to the sample and results are almost immediate. Light microscopy can require a first sorting of particles that can be operated by a sedimentation step. Proteomic approaches need a protein extraction step but can also be applied directly to the sample. Genomic methods require DNA extraction but this step can be semi-automated. A full automation would require having standardized samples that are not often the case in the feed sector due to the large variety of products. Genomic approaches demand management of the risk of contaminations (a single DNA molecule can provide billions of copies after PCR amplification) and involve the use of a high amount of non-reusable plastic wares. This last characteristic makes the genomic approaches not the most environmental-friendly with the concern to move in the direction of greener technologies. Spectroscopy is unquestionably the greenest technology as a very low quantity of specific reagents is needed (only during the training stage of the sensor) while other technologies need solvents that can require a specific treatment in some cases.

All technologies need top-level equipment and the level of sophistication is endlessly growing. Genomic methods have the highest price per sample due to the need for specialized reagents for DNA extraction, PCR amplification or sequencing.

Light microscopy, vibrational spectroscopy and omics approaches have a common point: the quantity of data provided has drastically increased and needs to be analyzed computationally. In general, the process is identical whatever the technology used. First, we have data acquisition, with an increasing flow and amount of data. Secondly, the data treatment is specific to each technology, for example, 'spectral' data are not analyzed in the same way as 'sequences'. Thirdly, the extraction and export of interesting information allows a decision to be made on the product. The creation of reference databases (spectral, imaging, DNA and protein sequences) is a crucial point to propose accurate values or classifications.

Data science is a generic word that can be applied to all fields of sciences (Sanchez-Pinto et al., 2018). Chemometrics became an important part of spectroscopy to solve predictive and classification problems (Fernández Pierna et al., 2012; Granato et al., 2018) but is also applied to other technologies such as proteomics (Lu et al., 2017). Automated image processing and analysis solutions are used for light microscopy (Arganda-Carreras and Andrey, 2017). Bioinformatics is expanding with HTS and mass spectrometry technologies (Kumar and Mann, 2009; Magi et al., 2010; Debode et al., 2019). Open-source programming and scripting languages have democratized access to data science techniques and statistical approaches (Sanchez-Pinto et al., 2018). Analysis can be fully automated even if we have to consider bioinformatic pipelines as dynamic systems that will evolve as a function of the knowledge and of the packages that will be available for the scientific community.

With the explosion of data provided, expertise in data science is required to properly implement all the technologies. Skills for processing the huge volume of data in the area of big data analysis will need to be intensified in the future. Such talents are much in demand and must often be developed internally in the research group or control laboratories. They will be used for a specific technology, database management and the development of several levels of analysis from the untargeted approach to specific detection. A future challenge for food and feed analysis will be to develop Artificial Intelligence and to apply it to machine learning so machines will be able to learn from previous data sets and make decisions by themselves. Meta-analysis of data coming from different approaches (through data merging, for instance) and handled together could also provide more complete information on the product or help to make a data-driven decision.

Another expected evolution is the migration of a part of the analysis from the lab to the field or the industry (lab-to-sample approach). Intense research and industrial initiatives are being performed in the area of portable and online solutions. Stand-alone sensors and mobile labs are in development to carry out an initial screening in the field, at the reception stage or in the industry. More precise or a confirmatory analysis could be mobilized in a second step in a classical laboratory. Smart solutions are also possible with a collection data with a smart instrumentation (even coupled to a smartphone) and sending the information in the cloud for treatment.

Danezis et al. (2016) observed that the published research in food authentication is exponential, and we can expect the same trends in feed. New feed products such as insect meal are also arriving on the market and need new analytical methods based on innovative technologies to address the quality, safety and authentication of the products.

8 Where to look for further information

Further information on the official control methods (light microscopy and PCR) and methods in development (mass spectrometry) for the detection of processed animal proteins can be found on the EURL-AP website (www.eurl.craw.eu).

For spectroscopic methods and chemometric tools, more information can be found in a large number of published books, as the Comprehensive Chemometrics book (Brown et al., 2020) or the Handbook of Chemometrics and Qualimetrics (Massart et al., 1997) book as well as dedicated trainings. CRA-W organizes an annual training on Vibrational Spectroscopy and Chemometrics (13th edition in 2020).

Concerning genomic methods, the European Network of GMO Laboratories (ENGL) plays an important role in the development, harmonization and standardization of methods from classic and emerging technologies. Their activities are mainly focused on GMO detection but can be transposed to other fields. Many of their documents can be found on the website of the European Union Reference Laboratory for GM Food and Feed (EURL-GMFF – https://gmo-crl.jrc.ec.europa.eu).

The developments of analytical methods in food and feed are periodically presented in dedicated conferences such as 'Recent Advances in Food Analysis' (RAFA – www.rafa2019.eu) and 'Rapid Methods Europe' (RME – www.rapidmethods.eu).

9 References

Abbas, O., Fernández Pierna, J. A., Boix, A., von Holst, C., Dardenne, P. and Baeten, V. 2010. Key parameters for the development of a NIR microscopic method for the quantification of by-products of animal origin in compound feedingstuffs. *Anal. Bioanal. Chem.* 397(5): 1965–1973.

Abbas, O., Fernández Pierna, J. A., Codony, R., von Holst, C. and Baeten, V. 2009. Assessment of the discrimination of animal fat by FT-Raman Spectroscopy. *J. Mol. Struct.* 924–926: 294–300.

Abbas, O., Lecler, B., Dardenne, P. and Baeten, V. 2013. Detection of melamine in feed ingredients by near infrared spectroscopy and chemometrics. *J. Near Infrared Spec.* 21(3): 183–194.

Aboling, S., Rieger, H., Kölln, M., Tenhündfeld, J., Roerink, G., Platje, N. and Kamphues, J. 2019. Feed refusal in fattening bulls because of maize silage contamination by jimson weed (Datura stramonium). *Tierarztl Prax. Ausg. G. Grosstiere Nutztiere* 47(2): 125–130.

Acunha, T., Ibáñez, C., García-Cañas, V., Simo, C. and Cifuentes, A. 2016. Recent advances in the application of capillary electromigration methods for food analysis and Foodomics. *Electrophoresis* 37(1): 111–141.

Álvarez-Rivera, G., Cifuentes, A. and Puyana, M. C. 2018. Electrophoretic technique: capillary zone electrophoresis. In: *Modern Techniques for Food Authentication*, 659–685 Sun, D.-W. (Ed.). Academic Press. 777 p. ISBN: 9780128142646.

Andueza, D., Picard, F. and Aufrère, J. 2013. Capability of local and global fresh permanent grassland calibrations to predict the chemical composition and nutritive value of permanent grassland hays using NIRS. In: 16th International Conference on Near Infrared Spectroscopy 2-7 June 2013, la Grande-Motte, France, 269–273.

Andueza, D., Picard, F., Jestin, M., Andrieu, J. and Baumont, R. 2011. NIRS prediction of the feed value of temperate forages: efficacy of four calibration strategies. *Animal* 5(7): 1002–1013.

Arganda-Carreras, I. and Andrey, P. 2017. Designing image analysis pipelines in light microscopy: a rational approach. *Methods Mol Biol.* 1563: 185–207.

Arulandhu, A. J., van Dijk, J., Staats, M., Hagelaar, R., Voorhuijzen, M., Molenaar, B., van Hoof, R., Li, R., Yang, L., Shi, J., Scholtens, I. and Kok, E. 2018. NGS-based amplicon sequencing approach; towards a new era in GMO screening and detection. *Food Control* 93: 201–210.

Axmann, S., Adler, A., Brandstettner, A. J., Spadinger, G., Weiss, R. and Strnad, I. 2015. Species identification of processed animal proteins (PAPs) in animal feed containing feed materials from animal origin. *Food Addit. Contam.* 32(7): 1089–1098.

Aydin, Ö. D., Yildiz, G., Güntürkün, O. B. and Bayraktaroglu, A. G. 2020. The use of glauconite as a feed additive in broiler nutrition and its effect on growth performance, intestinal histomorphology and biomechanical properties of bones. *Kafkas Univ. Vet. Fak. Derg.* 26(3): 343–349.

Baek, S. H., Lim, H. B. and Chun, H. S. 2014. Detection of melamine in foods using Terahertz Time-Domain spectroscopy. *J. Agric. Food Chem.* 62(24): 5403–5407.

Baeten, V., Meurens, M., Morales, M. T. and Aparicio, R. 1996. Detection of virgin olive oil adulteration by Fourier transform Raman spectroscopy. *J. Agric. Food Chem.* 44(8): 2225–2230.

Baeten, V., von Holst, C., Fissiaux, I., Michotte Renier, A., Murray, I. and Dardenne, P. 2004. The near infrared microscopic (NIRM) method: combination of the advantages of optical microscopy; near-infrared spectroscopy (WP5-NIRM). In: *Strategies; Methods to Detect; Quantify Mammalian Tissues in Feedingstuffs*. Bruxelles - Belgium, European Commission. ISBN: 9289473568.

Baeten, V., von Holst, C., Garrido, A., VanCutsem, J., Michotte Renier, A. and Dardenne, P. 2005. Detection of banned meat; bone meal in feedingstuffs by near-infrared microscopy analysis of the sediment fraction. *Anal. Bioanal. Chem.* 382(1): 149–157.

Baeten, V., Fernández Pierna, J. A. and Dardenne, P. 2007. Hyperspectral imaging techniques: an attractive solution for the analysis of biological and agricultural materials. In: *Techniques and Applications of Hyperspectral Image Analysis* Grahn, H. F. and Geladi, P. (Eds), 289–311. John Wiley & Sons, Ltd. ISBN: 9780470010860.

Baeten, V., Fernández Pierna, J. A., Dehareng, F., Sinnaeve, G. and Dardenne, P. 2010. Regulatory considerations in applying vibrational spectroscopic methods for quality control. In: *Applications of Vibrational Spectroscopy in Food Science*, 595–607 Li-Chan E.C.Y. (Ed.). John Wiley & Sons, Ltd. ISBN: 9780470742990.

Baeten, V. and Dardenne, P. 2016. NIR-Based detection of contaminants in food and feed. *Feedipedia* 35: 1–4.

Baeten, V., Pierna, J. A. F., Lecler, B., Abbas, O., Vincke, D., Minet, O., Vermeulen, P. and Dardenne, P. 2016. Near infrared spectroscopy for food and feed: a mature technique. *NIR news.* 27(1): 4–6.

Banati, D. 2011. Consumer response to food scandals and scares. *Trends Food Sci. Technol.* 22(2-3): 56–60.

Barbosa, C., Nogueira, S., Gadanho, M. and Chaves, S. 2016. Chapter 7. DNA extraction: finding the most suitable method. In: *Molecular Microbial Diagnostic Methods*, 135-154. Cook, N., D'Agostino, M. and Thompson K.C. (Eds). Academic Press, San Diego. ISBN: 9780124169999.

Barbosa, C., Nogueira, S., Saraiva, R. and Chaves, S. 2017. Next generation sequencing in food authenticity and safety. In: *Food Traceability and Authenticity*, 164–179 Montet D. and Ray R.C. (Eds). CRC Press. eBook ISBN: 9781351228435.

Bassi, D., Fontana, C., Zucchelli, S., Gazzola, S. and Cocconcelli, P. S. 2013. TaqMan Real time-quantitative PCR targeting the phosphotransacetylase gene for Clostridium tyrobutyricum quantification in animal feed, faeces, milk and cheese. *Int. Dairy J.* 33(1): 75–82.

Bassols, A., Turk, R. and Roncada, P. 2014. A proteomics perspective: from animal welfare to food safety. *Curr. Protein Pept. Sci.* 15(2): 156–168.

Belghit, I., Lock, E. J., Fumière, O., Lecrenier, M. C., Renard, P., Dieu, M., Berntssen, M. H. G., Palmblad, M. and Rasinger, J. D. 2019. Species-specific discrimination of insect meals for aquafeeds by direct comparison of tandem mass spectra. *Animals* 9(5): 222.

Belinchón Crespo, C., Veys, P., Vermeulen, P. and Baeten, V. 2012. Definition of key parameters for constructing an online micrograph collection of processed animal particles in feed. *Biotechnol. Agron. Soc. Environ.* 16(4): 452–462.

Boix, A., Fernández Pierna, J. A., von Holst, C. and Baeten, V. 2012. Validation of a near infrared microscopy method for the detection of animal products in feedingstuffs: results of a collaborative study. *Food Addit. Contam. A* 29(12): 1872 -1880.

Borràs, E., Ferré, J., Boqué, R., Mestres, M., Aceña, L. and Busto, O. 2015. Data fusion methodologies for food and beverage authentication and quality assessment–A review. *Analytica Chimica Acta* 891: 1–14.

Boval, M., Coates, D. B., Lecomte, P., Decruyenaere, V. and Archimède, H. 2004. Faecal Near Infrared Reflectance Spectroscopy (NIRS) to assess chemical composition, in vivo digestibility and intake of tropical grass by Creole cattle. *Anim. Feed Sci. Technol.* 114(1-4): 19–29.

Brera, C., De Santis, B. and Prantera, E. 2012. Animal feed sampling for contaminant analysis. In: *Animal Feed Contamination Effects on Livestock and Food Safety* Fink-Gremmels, J. (Ed), 557-588. Cambridge, United Kingdom, Woodhead Publishing. ISBN: 9781845697259.

Broeders, S., Huber, I., Grohmann, L., Berben, G., Taverniers, I., Mazzara, M., Roosens, N. and Morisset, D. 2014. Guidelines for validation of qualitative real-time PCR methods. *Trends Food Sci. Technol.* 37(2): 115–126.

Brown, S., Tauler, R. and Walczak, B. (Eds). 2020. *Comprehensive Chemometrics: Chemical and Biochemical Data Analysis*. Elsevier, 2944 p. ISBN: 9780444641656.

Bryden, W. L. 2012. Mycotoxin contamination of the feed supply chain: implications for animal productivity and feed security. *Anim. Feed Sci. Tech.* 173(1-2): 134–158.

Buckley, M., Collins, M., Thomas-Oates, J. and Wilson, J. C. 2009. Species identification by analysis of bone collagen using matrix-assisted laser desorption/ionisation time-of-flight mass spectrometry. *Rapid Commun. Mass Spectrom.* 23(23): 3843-3854.

Bultreys, A. and Gheysen, I. 2020. First report of *Xanthomonas phaseoli* pv. *phaseoli* and *Xanthomonas citri* pv. *fuscans* causing common bacterial blight of bean in Belgium. *New Disease Reports* 41: 6.

Cai, Y., He, Y., Lv, R., Chen, H., Wang, Q. and Pan, L. 2017. Detection and quantification of beef and pork materials in meat products by duplex droplet digital PCR. *PLoS ONE* 12(8): e0181949.

Campos, M. D., Valadas, V., Campos, C., Morello, L., Braglia, L., Breviario, D. and Cardoso, H. G. 2018. A TaqMan Real-time PCR method based on alternative oxidase genes for detection of plant species in animal feed samples. *PLoS ONE* 13(1): e0190668.

Caporaso, N., Whitworth, M. B. and Fisk, I. D. 2018. Near-infrared spectroscopy and hyperspectral imaging for non-destructive quality assessment of cereal grains. *Appl. Spectrosc. Rev.* 53(8): 667–687.

Capuano, E., Boerrigter-Eenling, R., Koot, A. and van Ruth, S. M. 2015. Targeted and untargeted detection of skim milk powder adulteration by near-infrared spectroscopy. *Food analytical methods* 8(8): 2125–2134.

Carvalho, D. C., Palhares, R. M., Drummond, M. G. and Gadanho, M. 2017. Food metagenomics: next generation sequencing identifies species mixtures and mislabeling within highly processed cod products. *Food Control* 80: 183–186.

Chu, X., Wang, W., Yoon, S. C., Ni, X. and Heitschmidt, G. W. 2017. Detection of aflatoxin b1 (AFB1) in individual maize kernels using short wave infrared (SWIR) hyperspectral imaging. *Biosyst. Eng.* 157: 13–23.

Clerens, S., Plowman, J. E. and Dyer, J. M. 2012. Food proteomics: mapping modifications. In: *Proteomic Applications in Biology* Heazlewood, J. L. and Petzold, C. J. (Eds), 3–32. IntechOpen. doi: 10.5772/31308.

Colgrave, M. L. 2017. *Proteomics in Food Science: From Farm to Fork*. Academic Press. doi: 10.1016/B978-0-12-804007-2.05001-3. ISBN: 9780128040072.

Colovic, D., Kokic, B., Cabarkapa, I., Levic, J., Djuragic, O., Teichmann, K. and Jedrejek, D. 2014. Hyperspectral Imaging in plant based food safety. Possibility of hyperspectral Imaging application in safety control of plant based food that is converted to by-products and used as feed. *Agro Food Ind. Hi Tech* 25(5): 39–43.

Cozzolino, D. and Labandera, M. 2002. Determination of dry matter and crude protein contents of undried forages by near-infrared reflectance spectroscopy. *J. Sci. Food Agric.* 82(4): 380–384.

Croote, D. and Quake, S. R. 2016. Food allergen detection by mass spectrometry: the role of systems biology. *NPJ Syst. Biol. Appl.* 2: 16022.

Cunsolo, V., Muccilli, V., Saletti, R. and Foti, S. 2014. Mass spectrometry in food proteomics: a tutorial. *J. Mass Spectrom.* 49(9): 768–784.

Dale, L. M., Thewis, A., Boudry, C., Rotar, I., Dardenne, P., Baeten, V. and Fernández Pierna, J. A. F. 2013. Hyperspectral imaging applications in agriculture and agro-food product quality and safety control: a review. *Appl. Spectrosc. Rev.* 48(2): 142–159.

Dale, L. M., Thewis, A., Rotar, I., Boudry, C., Vidican, R., Malinas, A., Florian, V., Lecler, B., Agneessens, R., Fernández Pierna, J. A. and Baeten, V. 2012. Determination of Romanian alfalfa crude protein and crude fiber contents as well as in vitro organic matter digestibility by NIR spectrometry. *Manag. Econ. Eng. Agric. Rur. Develop* 12(4): 57–62.

Danezis, G. P., Tsagkaris, A. S., Camin, F., Brusic, V. and Georgiou, C. A. 2016. Food authentication: Techniques, trends & emerging approaches. *Trends in Analytical Chemistry* 85: 123–132.

D'Alessandro, A. and Zolla, L. 2012. We are what we eat: food safety and proteomics. *J. Proteome Res.* 11(1): 26–36.

Dangel, A., Berger, A., Messelhäußer, U., Konrad, R., Hörmansdorfer, S., Ackermann, N. and Sing, A. 2019. Genetic diversity and delineation of Salmonella Agona outbreak strains by next generation sequencing, Bavaria, Germany, 1993 to 2018. *EURO Surveill.* 24(18): 31064635.

Davis, A. N., McIntyre, P. and Morgan, E. 2000. Study of the use of molecular spectroscopy for the authentication of extra virgin olive oils, part I: fourier transform Raman spectroscopy. *Appl. Spectrosc.* 54(12): 1864–1867.

De Battisti, C., Marciano, S., Magnabosco, C., Busato, S., Arcangeli, G. and Cattoli, G. 2014. Pyrosequencing as a tool for rapid fish species identification and commercial fraud detection. *J. Agr. Food Chem.* 62(1): 198–205.

Debode, F., Huber, I., Macarthur, R., Rischitor, P. E., Mazzara, M., Herau, V., Sebah, D., Dobnik, D., Broeders, S., Roosens, N. H., Bush, U., Berben, G., Morisset, D. and Zel, J. 2017c. Inter-laboratory studies for the validation of two singleplex (tE9 and pea lectin) and one duplex (pat/bar) real-time PCR methods for GMO detection. *Food Control* 73: 452–461.

Debode, F., Hulin, J., Charloteaux, B., Coppieters, W., Hanikenne, M., Karim, L. and Berben, G. 2019. Detection and identification of transgenic events by next generation sequencing combined with enrichment technologies. *Sci. Rep.* 9(1): 15595.

Debode, F., Janssen, E., Bragard, C. and Berben, G. 2017d. Detection by real-time PCR and Pyrosequencing of the cry 1Ab and cry 1Ac genes introduced in genetically modified (GM) constructs. *Food Addit. Contam. A* 34(8): 1398–1409.

Debode, F., Janssen, E. and Berben, G. 2007. Physical degradation of genomic DNA of soybean flours does not impair relative quantification of its transgenic content. *Eur. Food Res. Technol.* 226(1-2): 273–280.

Debode, F., Janssen, E. and Berben, G. 2013. Development of 10 new screening PCR assays for GMO detection targeting promoters (pFMV, pNOS, pSSuAra, pTA29, pUbi, pRice actin) and terminators (t35S, tE9, tOCS, tg7). *Eur. Food Res. Technol.* 236(4): 659–669.

Debode, F., Janssen, E. and Berben, G. 2018b. Development of PCR screening assays focused on gene-coding sequences for GMO detection. *Base* 22(4): 1–12.

Debode, F., Marien, A., Gérard, A., Francis, F., Fumière, O. and Berben, G. 2017. Development of real-time PCR tests for the detection of Tenebrio molitor in food and feed. *Food Addit. Contam. A* 34(8): 1421–1426.

Debode, F., Marien, A., Janssen, E., Bragard, C. and Berben, G. 2017b. The influence of amplicon length on real-time PCR results. *Base* 21(1): 3–11.

Debode, F., Zdeňková, K., Janssen, E., Tizolova, A., du Jardin, P., Berben, G. and Demnerova, K. 2018. Development of real-time PCR assays for the detection of the pin II terminator (tpinII) used in GM constructs and its donor organism, potato (Solanum tuberosum). *Food Anal. Methods* 11(8): 2172–2180.

Decruyenaere, V. 2015. Estimation of diet digestibility and intake of grazing ruminants through near infrared reflectance spectroscopy analysis of faeces. Application in various contexts of livestock production. (PhD Dissertation in English). Gembloux Agro-Bio Tech, University of Liège and Gembloux. Belgium, p. 163. Available at: https://orbi.uliege.be/bitstream/2268/189025/1/phD%20VDecruyenaere-final-20151209.pdf (Last accessed 20/04/2020).

Dee, S. A. , Bauermann, F. V., Niederwerder, M. C., Singrey, A., Clement, T., de Lima, M., Long, C., Patterson, G., Sheahan, M. A., Stoian, A. M. M., Petrovan, V., Jones, C. K., De Jong, J., Ji, J., Spronk, G. D., Minion, L., Christopher-Hennings, J., Zimmerman,

J. J., Rowland, R. R. R., Nelson, E., Sundberg, P., Diel, D. G. 2018. Survival of viral pathogens in animal feed ingredients under transboundary shipping models. *PLoS ONE* 13(3). Available at: e0194509.

De la Haba, M. J., Fernández Pierna, J. A., Fumière, O., Garrido-Varo, A., Guerrero, J. E., Pérez-Marín, D. C., Dardenne, P. and Baeten, V. 2007. Discrimination of fish bones from other animal bones in the sedimented fraction of compound feeds by near infrared microscopy (NIRM). *J. NIRS* 15: 81–88.

Demeke, T. and Jenkins, G. R. 2010. Influence of DNA extraction methods, PCR inhibitors and quantification methods on real-time PCR assay of biotechnology-derived traits. *Anal. Bioanal. Chem.* 396(6): 1977–1990.

Directive 2002/32/EC of the European Parliament and of the Council of 7 May 2002 on undesirable substances in animal feed. *Off. J. Eur. Union L.* 140: 10–21.

Dobnik, D., Štebih, D., Blejec, A., Morisset, D. and Žel, J. 2016. Multiplex quantification of four DNA targets in one reaction with Bio-Rad droplet digital PCR system for GMO detection. *Sci. Rep.* 6: 35451.

Domesle, K. J., Yang, Q., Hammack, T. S. and Ge, B. 2018. Validation of a Salmonella loop-mediated isothermal amplification assay in animal food. *Int. J.Food Microbiol.* 264: 63–76.

Dong, Z., Li, J., Chen, L., Wang, S., Liu, Q. and Shao, T. 2019. Using proteomics to decipher the effect of tissue damage on the fate of red clover proteins during ensiling. *Anim. Feed Sci. Technol.* 257: 114271.

Donnelly, D. M., Dórea, J. R. R., Yang, H. and Combs, D. K. 2018. Technical note: comparison of dry matter measurements from handheld near-infrared units with oven drying at 60°C for 48 hours and other on-farm methods. *J. Dairy Sci.* 101(11): 9971–9977.

Dumont, V., Kerbach, S., Poms, R., Johnson, P., Mills, C., Popping, B., Tömösközi, S. and Delahaut, P. 2010. Development of milk and egg incurred reference materials for the validation of food allergen detection methods. *Qual. Assur. Saf. Crop. Foods* 2: 208–215.

EC Regulation 178/2002 of the European Parliament and of the Council of 28 January 2002 laying down the general principles and requirements of food law, establishing the European Food Safety Authority and laying down procedures in matters of food safety. *Off. J. Eur. Union L.* 031: 01–24.

EC Regulation 152/2009 of the European Parliament and of the Council of 27 January 2009 laying down the methods of sampling and analysis for the official control of feed. *Off. J. Eur. Union L.* 54: 1–130.

EC Regulation 51/2013 of the European Parliament and of the Council of 16 January 2013 amending Regulation (EC) no 152/2009 as regards the methods of analysis for the determination of constituents of animal origin for the official control of feed. *Off. J. Eur. Union L.* 20: 33–43.

EC Regulation 56/2013 of the European Parliament and of the Council of 16 January 2013 amending Annexes I and IV to Regulation (EC) No 999/2001 of the European Parliament and of the Council laying down rules for the prevention, control and eradication of certain transmissible spongiform encephalopathies. *Off. J. Eur. Union L* 21: 3–16.

EFSA, European Food Safety Authority. 2018. Outcome of EC/EFSA questionnaire (2016) on use of Whole Genome Sequencing (WGS) for food-and waterborne pathogens isolated from animals, food, feed and related environmental samples in EU/EFTA countries. *EFSA Support. Publ.* 15(6): 1432E.

Ehlers, S. M., Maxein, J. and Koop, J. H. E. 2020. Low-cost microplastic visualization in feeding experiments using an ultraviolet light-emitting flashlight. *Ecol. Res.* 35(1): 265–273.

ENGL. 2015. *Definition of Minimum Performance Requirements for Analytical Methods for GMO Testing*. Report number: JRC95544. Ispra, Italy, European Commission–Joint Research Centre (JRC). doi: 10.13140/RG.2.1.2060.5608.

ENGL. 2017. *Verification of analytical methods for GMO testing when implementing interlaboratory validated methods. Version 2*. Report number: EUR29015. Ispra, Italy, European Commission–Joint Research Centre (JRC). doi: 10.2760/645114.

Erickson, B. E. 1999. Focus: dioxin food crisis in Belgium. *Anal. Chem.* 71(15): 541A-543A.

Fakruddin, M. D. and Chowdhury, A. 2012. Pyrosequencing-an alternative to traditional Sanger sequencing. *Am. J. Biochem. Biotechnol.* 8(1): 14–20.

FAO/WHO Expert Meeting on Hazards Associated with Animal Feed 12-15 May 2015 FAO Headquarters, Rome, Italy Executive Summary of the Report www.fao.org/3/a-az851e.pdf.

Feider, C. L., Krieger, A., DeHoog, R. J. and Eberlin, L. S. 2019. Ambient ionization mass spectrometry: recent developments and applications. *Anal. Chem.* 91(7): 4266–4290.

FEFAC. 2016. Fefac vision on animal feed industry. Available at: https://www.fefac.eu/files/67546.pdf (Last accessed 20/04/2020).

Fernández-Carballo, B. L., McGuiness, I., McBeth, C., Kalashnikov, M., Borrós, S., Sharon, A. and Sauer-Budge, A. F. 2016. Low-cost, real-time, continuous flow PCR system for pathogen detection. *Biomed. Microdevices* 18(2): 34.

Fernández Pierna, J. A., Boix, A., Slowikowski, B., von Holst, C., Maute, O., Han, L., Amato, G., de la Roza Delgado, B., Marín, D., Lilley, G., Dardenne, P. and Baeten, V. 2013. Standardization of NIR microscopy spectra obtained from inter-laboratory studies by using a standardization cell. *Base* 17(4): 547–555.

Fernández Pierna, J. A., Dardenne, P. and Baeten, V. 2010b. In-house validation of a near infrared hyperspectral imaging method for detecting processed animal proteins (PAP) in compound feed. *J. NIRS* 18: 121-133.

Fernández Pierna, J. A., Vermeulen, P., Amand, O., Tossens, A., Dardenne, P. and Baeten, V. 2012. NIR hyperspectral imaging spectroscopy and chemometrics for the detection of undesirable substances in food and feed. *La.B.* 117: 233–239.

Fernández Pierna, J. A., Vermeulen, P., Eylenbosch, D., Burger, J., Bodson, B., Dardenne, P. and Baeten, V. 2020. Chemometrics in NIR hyperspectral Imaging–theory and applications in the agricultural crops and products sector. In: *Comprehensive Chemometrics* Brown, S., Tauler, R. and Walczak, B. (Eds), (Vol. 4). Oxford and Elsevier. doi: 10.1016/B978-0-12-409547-2.14811-5.

Fernández Pierna, J. A., Vermeulen, P., Lecler, B., Baeten, V. and Dardenne, P. 2010a. Calibration transfer from dispersive instruments to handheld spectrometers. *Appl. Spectrosc.* 64(6): 644–648.

Fibi, S., Klose, V., Mohnl, M., Weber, B., Haslberger, A. G. and Sattler, V. A. 2016. Suppression subtractive hybridisation and real-time PCR for strain-specific quantification of the probiotic Bifidobacterium animalis BAN in broiler feed. *J.Microbiol. Methods* 123: 94-100.

Fraiture, M. A., Saltykova, A., Hoffman, S., Winand, R., Deforce, D., Vanneste, K., De Keersmaecker, S. C. J., Roosens, N. H. C. 2018. Nanopore sequencing technology: a new route for the fast detection of unauthorized GMO. *Sci. Rep.* 8(1): 7903.

Francis, F., Mazzucchelli, G., Baiwir, D., Debode, F., Berben, G. and Caparros Megido, R. 2020. Proteomics based approach for edible insect fingerprinting in novel food: differential efficiency according to selected model species. *Food Control* 112: 107135.

Frick, G., Boschung, H., Schulz-Schroeder, G., Russ, G., Ujčič-Vrhovnik, I., Jakovac-Strajn, B., Angetter, D., John, I. and Jørgensen, J. S. 2011. Ragweed (*Ambrosia* sp.) seeds in bird feed. *Biotechnol. Agron. Soc. Environ.* 15 (Suppl.1): 34–44.

Fumière, O., Dubois, M., Baeten, V., von Holst, C. and Berben, G. 2006. Effective PCR detection of animal species in highly processed animal byproducts and compound feeds. *Anal. Bioanal. Chem.* 385(6) : 1045–1054.

Fumière, O., Marien, A., Fernández Pierna, J. A., Baeten, V. and Berben, G. 2008. Development of an original DNA extraction protocol for the species specific identification of PAPs particles in feeds using NIRM and Real Time PCR. In: Proceedings of the Rapid Methods Europe 2008 for food and feed safety and quality, January 21-23, 2008. Europe. Noordwijkerhout, The Netherlands. Bilthoven, The Netherlands, Rapid Methods. Available at: https://www.bastiaanse-communication.com/rme2013/boa8.pdf (Last accessed 28/04/2020).

Fumière, O., Marien, A., Fernández Pierna, J. A., Baeten, V. and Berben, G. 2010. Development of a real-time PCR protocol for the species origin confirmation of isolated animal particles detected by NIRM. *Food. Addit. Contam. A* 27(8): 1118–1127.

Furst, P. 2011. Dioxins in feed and food again-real or perceived risk? *Eur. J. Lipid Sci. Technol.* 113(4): 401–402.

Furutani, S., Hagihara, Y. and Nagai, H. 2017. On-site identification of meat species in processed foods by a rapid real-time polymerase chain reaction system. *Meat Sci.* 131: 56–59.

Garrido, A., Chapela, M. J., Román, B., Ferreira, M., Lago, J., Vieites, J. M. and Cabado, A. G. 2012. Development of a multiplex real-time PCR method for simultaneous detection of Salmonella enterica, Shigella flexneri and Listeria monocytogenes in processed food samples. *European Food Research and Technology* 234(4): 571–580.

Garrido, A., Chapela, M. J., Román, B., Fajardo, P., Lago, J., Vieites, J. M. and Cabado, A. G. 2013. A new multiplex real-time PCR developed method for Salmonella spp. and Listeria monocytogenes detection in food and environmental samples. *Food Control* 30(1): 76–85.

Gassner, G. 1931. *Mikroskopische Untersuchung pflanzlicher Nahrungs- und Genussmittel*. Jena, Germany, Gustav Fischer. ISBN: 9783899472561.

Golinelli, L. P., Silva, J. T., Carvalho, A. C. and Paschoalin, V. M. F. 2016. Detection of animal products in ruminant feeds by microscopy and real time PCR. *J.Vet. Sci. Technol.* 7: 316.

Gomez-Morte, T., Ayala-Hernández, M., Yánez-Gascón, M. J., Gil-Izquierdo, A., Auñon-Calles, D. A., Domínguez-Perles, R., Fortea, M. I., Núñez-Delicado, E. and Gabaldón, J. A. 2019. Immunoassay for food quality evaluation. In: *Evaluation Technologies for Food Quality* Zhong, J. and Wang, X. (Eds), 661-695. Woodhead Publishing. doi: 10.1016/B978-0-12-814217-2.00026-3.

Gou, T., Hu, J., Wu, W., Ding, X., Zhou, S., Fang, W. and Mu, Y. 2018. Smartphone-based mobile digital PCR device for DNA quantitative analysis with high accuracy. *Biosens. Bioelectron.* 120: 144–152.

Gowen, A. A., O'Donnell, C. P., Cullen, P. J., Downey, G. and Frias, J. M. 2007. Hyperspectral imaging–an emerging process analytical tool for food quality and safety control. *Trends Food Sci.Technol.* 18(12): 590–598.

Graham, S. F., Haughey, S. A., Ervin, R. M., Cancouët, E., Bell, S. and Elliott, C. T. 2012. The application of near-infrared (NIR) and Raman spectroscopy to detect adulteration of oil used in animal feed production. *Food Chem.* 132(3): 1614–1619.

Granato, D., Putnik, P., Kovačević, D. B., Santos, J. S., Calado, V., Rocha, R. S., Gomes Da Cruz, A. G. D., Jarvis, B., Rodionova, O. Y. and Pomerantsev, A. 2018. Trends in chemometrics: food authentication, microbiology, and effects of processing. *Compr. Rev. Food Sci. Food Saf.* 17(3): 663–677.

Hahn, H. and Michaelsen, I. 1996. *Mikroskopische Diagnostik pflanzlicher Nahrungs-, Genuss- und Futtermittel, einschliesslich Gewürze*. Berlin, Germany, Springer-Verlag. ISBN: 9783540600527.

Hart, E. H., Onime, L. A., Davies, T. E., Morphew, R. M. and Kingston-Smith, A. H. 2016. The effects of PPO activity on the proteome of ingested red clover and implications for improving the nutrition of grazing cattle. *J. Proteomics* 141: 67–76.

Haynes, E., Jimenez, E., Pardo, M. A. and Helyar, S. J. 2019. The future of NGS (Next Generation Sequencing) analysis in testing food authenticity. *Food Control* 101: 134–143.

Higuchi, R., Fockler, C., Dollinger, G. and Watson, R. 1993. Kinetic PCR analysis: real-time monitoring of DNA amplification reactions. *Nat. Biotechnol.* 11(9): 1026–1030.

Hohmann, B. and Gassner, G. 2007. *Mikroskopische Untersuchung pflanzlicher Lebensmittel und Futtermittel: der Gassner* (6th edn.). Hamburg, Germany, Behr's Verlag. ISBN: 9783899472561.

Holland, P. M., Abramson, R. D., Watson, R. and Gelfand, D. H. 1991. Detection of specific polymerase chain reaction product by utilizing the 5'----3'exonuclease activity of Thermus aquaticus DNA polymerase. *Proc. Natl Acad. Sci. U.S.A.* 88(16): 7276–7280.

Horn, G., Bobrow, M., Bruce, M., Goedert, M., McLean, A. and Webster, J. 2001. Review of the origin of BSE: report to United Kingdom Government. (p. 47.46). Available at: http, 111 94–112; http://webarchive.nationalarchives.gov.uk/20090731155538 /http://www; defra.gov.uk/animalh/bse/publications/bseorigin.pdf (Last accessed 20/04/2020).

Hsu, C. C., Baker, M. W., Gaasterland, T., Meehan, M. J., Macagno, E. R. and Dorrestein, P. C. 2017. Top-down atmospheric ionization mass spectrometry microscopy combined With proteogenomics. *Anal. Chem.* 89(16): 8251–8258.

Huber, I., Block, A., Sebah, D., Debode, F., Morisset, D., Grohmann, L., Berben, G., Stebih, D., Milavec, M., Zel, J. and Busch, U. 2013. Development and validation of duplex, triplex, and pentaplex real-time PCR screening assays for the detection of genetically modified organisms in food and feed. *J. Agric. Food Chem.* 61(43): 10293–10301.

Jang, M., Jeong, S. W., Bae, N. H., Song, Y., Lee, T. J., Lee, M. K., Lee, S. J. and Lee, K. G. 2017. Droplet-based digital PCR system for detection of single-cell level of foodborne pathogens. *BioChip J.* 11(4): 329–337.

Jonkuviene, D., Salomskiene, J. and Zaborskiene, G. 2013. Rapid Pyrosequencing and fatty acid analysis for characterization of Bacillus cereus isolates from food. *Food Technol. Biotechnol.* 51(4): 564–569.

Jun-Shi, C. 2009. What can we learn from the 2008 melamine crisis in China? *Biomed. Environ. Sci.* 22(2): 109--111.

Kadhim, B. M. and Mater, H. N. 2019. Bacterial contaminants Associated with Poultry Feeds. *Ind. J. of Publ. Health Rese. & Develop.* 10(9): 964–969.

Kim, M. J. and Kim, H. Y. 2017. Species identification of commercial jerky products in food and feed using direct pentaplex PCR assay. *Food Control* 78: 1–6.

Kimuli, D., Wang, W., Lawrence, K. C., Yoon, S. C., Ni, X. and Heitschmidt, G. W. 2018b. Utilisation of visible/near-infrared hyperspectral images to classify aflatoxin b1 contaminated maize kernels. *Biosyst. Eng.* 166: 150–160.

Kimuli, D., Wang, W., Wang, W., Jiang, H., Zhao, X. and Chu, X. 2018a. Application of SWIR hyperspectral imaging and chemometrics for identification of aflatoxin b1 contaminated maize kernels. *Infrared Phys. Technol.* 89: 351-362.

Klein, H. and Marquard, R. 2005. *Feed Microscopy–Atlas for the Microscopic Examination of Feed Containing Vegetable and Animal Products*. Bergen, Germany, Agrimedia. ISBN: 9783860372555.

Köppel, R., Ganeshan, A., Weber, S., Pietsch, K., Graf, C., Hochegger, R., Griffiths, K. and Burkhardt, S. 2019. Duplex digital PCR for the determination of meat proportions of sausages containing meat from chicken, turkey, horse, cow, pig and sheep. *Eur. Food Res.Technol.* 245(4): 853–862.

Korte, R. and Brockmeyer, J. 2017. Novel mass spectrometry approaches in food proteomics. *TrAC Trends Anal. Chem.* 96: 99–106.

Košir, A. B., Demšar, T., Štebih, D., Žel, J. and Milavec, M. 2019. Digital PCR as an effective tool for GMO quantification in complex matrices. *Food Chem.* 294: 73–78.

Košir, A. B., Spilsberg, B., Holst-Jensen, A., Žel, J. and Dobnik, D. 2017. Development and inter-laboratory assessment of droplet digital PCR assays for multiplex quantification of 15 genetically modified soybean lines. *Sci. Rep.* 7(1): 8601.

Kumar, C. and Mann, M. 2009. Bioinformatics analysis of mass spectrometry-based proteomics data sets. *FEBS Let.* 583(11): 1703–1712.

Kumari, S., Kumar, R. R., Mendiratta, S. K., Kumar, D., Rana, P., Kumar, D. and Jawla, J. 2019. Species-specific loop-mediated isothermal amplification (LAMP) assay for identification of tissue of cattle origin by targeting mitochondrial gene sequences. *3 Biotech* 9(3): 69.

Kusebauch, U., Hernández-Castellano, L. E., Bislev, S. L., Moritz, R. L., Røntved, C. M. and Bendixen, E. 2018. Selected reaction monitoring mass spectrometry of mastitis milk reveals pathogen-specific regulation of bovine host response proteins. *J. Dairy Sci.* 101(7): 6532–6541.

Lecrenier, M. C. 2019. Development of complementary methods for the detection of animal by-products in feedingstuffs. Phd Thesis. Available at: http://hdl.handle.net /2268/235700 (Last accessed 28/04/2020).

Lecrenier, M. C., Ledoux, Q., Berben, G., Fumière, O., Saegerman, C., Baeten, V. and Veys, P. 2014. Determination of the ruminant origin of bone particles using fluorescence in situ hybridization (FISH). *Sci. Rep.* 4: 5730.

Lecrenier, M. C., Marbaix, H., Dieu, M., Veys, P., Saegerman, C., Raes, M. and Baeten, V. 2016. Identification of specific bovine blood biomarkers with a non-targeted approach using HPLC ESI tandem mass spectrometry. *Food Chem.* 213: 417–424.

Lecrenier, M. C., Planque, M., Dieu, M., Veys, P., Saegerman, C., Gillard, N. and Baeten, V. 2018. A mass spectrometry method for sensitive, specific and simultaneous detection of bovine blood meal, blood products and milk products in compound feed. *Food Chem.* 245: 981–988.

Lecrenier, M. C., Veys, P., Fumière, O., Berben, G., Saegerman, C. and Baeten, V. 2020. Official feed control linked to the detection of animal by-products: Past, present, and future. *Journal of Agricultural and Food Chemistry*, 68(31): 8093–8103.

Levin, R. E., Ekezie, F. G. C. and Sun, D. W. 2018. DNA-based technique: polymerase chain reaction (PCR). In: *Modern Techniques for Food Authentication*, 527–616. Academic Press. ISBN: 9780128142646.

Lian, D. S. and Zeng, H. S. 2017. Capillary electrophoresis based on nucleic acid detection as used in food analysis. *Compr. Rev. Food Sci. Food Saf.* 16(6): 1281–1295.

Li, J., Ding, Q., Zeng, X., Chen, K. and Zhou, L. 2018. Comparative analysis of real-time PCR and loop-mediated isothermal amplification in determination animal-origin ingredients in food. *J. Food Saf. Qual.* 9(12): 2908–2913.

Littlefair, J. E. and Clare, E. L. 2016. Barcoding the food chain: from Sanger to high-throughput sequencing. *Genome* 59(11): 946–958.

Liu, H., He, Y., Wang, M., He, W., Xie, R., Zheng, C. and Mo, L. 2018. Rapid detection of foodborne pathogenic bacteria by loop-mediated isothermal amplification. *J. Food Saf. Qual.* 9(16): 4412–4416.

Liu, W., Liu, B., Li, X., Li, H., Shi, H. and Wu, Y. 2017. Rapid detection of Listeria monocytogenes in food by real-time PCR. *J. Food Saf. Qual.* 8(1): 252–255.

Liu, Y., Chao, K., Kim, M. S., Tuschel, D., Olkhovyk, O. and Priore, R. J. 2009. Potential of Raman spectroscopy and imaging methods for rapid and routine screening of the presence of melamine in animal feed and foods. *Appl. Spectrosc.* 63(4): 477–480.

Lockley, A. K. and Bardsley, R. G. 2000. DNA-based methods for food authentication. *Trends Food Sci. Tech.* 11(2): 67–77.

Long, L. K., Yan, W., Li, C. C., Xia, W., Xing, Z. J., Liu, N., Dong, L. M. and Li, F. W. 2019. Rapid visual detection of four specific transgenic events in GM soybean using loop-mediated isothermal amplification method. *Russ. J. Plant Physiol.* 66(4): 646–655.

López-Díez, E. C., Bianchi, G. and Goodacre, R. 2003. Rapid quantitative assessment of the adulteration of virgin olive oils with hazelnut oils using Raman spectroscopy and chemometrics. *J. Agric. Food Chem.* 51(21): 6145–6150.

Lo, Y. T. and Shaw, P. C. 2018. DNA-based techniques for authentication of processed food and food supplements. *Food Chem.* 240: 767–774.

Lu, H., Zhang, H., Chingin, K., Xiong, J., Fang, X. and Chen, H. 2018. Ambient mass spectrometry for food science and industry. *TrAC, Trends Anal. Chem.* 107 : 99–115.

Lu, W., Liu, J., Gao, B., Lv, X. and Yu, L. L. 2017. Technical note: nontargeted detection of adulterated plant proteins in raw milk by UPLC-quadrupole time-of-flight mass spectrometric proteomics combined with chemometrics. *J. Dairy Sci.* 100(9): 6980–6986.

Magi, A., Benelli, M., Gozzini, A., Girolami, F., Torricelli, F. and Brandi, M. L. 2010. Bioinformatics for next generation sequencing data. *Genes* 1(2): 294–307.

Makowski, J., Vary, N., Mccutcheon, M. and Veys, P. 2010. *Microscopic Analysis of Agricultural Products*. Urbana, IL, AOCS Press. ISBN: 9781893997714.

Mandrile, L., Fusaro, I., Amato, G., Marchis, D., Martra, G. and Rossi, A. M. 2018. Detection of insect's meal in compound feed by Near infrared spectral imaging. *Food Chem.* 267(30): 240–245.

Marbaix, H., Budinger, D., Dieu, M., Fumière, O., Gillard, N., Delahaut, P., Mauro, S. and Raes, M. 2016. Identification of proteins and peptide biomarkers for detecting banned processed animal proteins (PAPs) in meat and bone meal by mass spectrometry. *J. Agric. Food Chem.* 64(11): 2405–2414.

Marchis, D., Altomare, A., Gili, M., Ostorero, F., Khadjavi, A., Corona, C., Ru, G., Cappelletti, B., Gianelli, S., Amadeo, F., Rumio, C., Carini, M., Aldini, G. and Casalone, C. 2017.

LC-MS/MS identification of species-specific muscle peptides in processed animal proteins. *J. Agric. Food Chem.* 65(48): 10638–10650.
Marco-Ramell, A., de Almeida, A. M., Cristobal, S., Rodrigues, P., Roncada, P. and Bassols, A. 2016. Proteomics and the search for welfare and stress biomarkers in animal production in the one-health context. *Mol. Biosyst.* 12(7): 2024–2035.
Maria John, K. M., Khan, F., Luthria, D. L., Garrett, W. and Natarajan, S. 2017. Proteomic analysis of anti-nutritional factors (ANF's) in soybean seeds as affected by environmental and genetic factors. *Food Chem.* 218: 321–329.
Martens, H. and Naes, T. 1989. *Multivariate Calibration* (2nd edn.) (Vol. 1). Chichester, UK, Wiley. ISBN: 0471909793.
Massart, D. L., Vandeginste, B. G. M., Buydens, L. M. C., De Jong, S., Lewi, P. J. and Smeyers-Verbeke, J. 1997. *Handbook of Chemometrics and Qualimetrics: Part A*, 886 p. Amsterdam, Elsevier. ISBN: 9780444897240.
Mayo, B., Rachid, C., Alegria, A., Leite, A., Peixoto, R. and Delgado, S. 2014. Impact of next generation sequencing techniques in food microbiology. *Curr. Genom.* 15(4): 293–309.
Michael, T. P. and Jackson, S. 2013. The first 50 plant genomes. *The plant genome*, 6(2). https://core.ac.uk/download/pdf/194193971.pdf.
Modroño, S., Soldado, A., Martínez-Fernández, A. and de la Roza-Delgado, B. 2017. Handheld NIRS sensors for routine compound feed quality control: real time analysis and field monitoring. *Talanta* 162: 597–603.
Montowska, M., Alexander, M. R., Tucker, G. A. and Barrett, D. A. 2014. Rapid detection of peptide markers for authentication purposes in raw and cooked meat using ambient liquid extraction surface analysis mass spectrometry. *Anal. Chem.* 86(20): 10257–10265.
Montowska, M., Alexander, M. R., Tucker, G. A. and Barrett, D. A. 2015. Authentication of processed meat products by peptidomic analysis using rapid ambient mass spectrometry. *Food Chem.* 187: 297–304.
Montowska, M. and Pospiech, E. 2013. Species-specific expression of various proteins in meat tissue: proteomic analysis of raw and cooked meat and meat products made from beef, pork and selected poultry species. *Food Chem.* 136(3–4): 1461–1469.
Morales Sillero, A., Fernández Pierna, J. A., Sinnaeve, G., Dardenne, P. and Baeten, V. 2018. Quantification of protein in wheat using NIR hyperspectral imaging: performance comparison with conventional NIR spectroscopy. *J.Near Infrared Spectrosc.* 26(3): 186–195.
Morisset, D., Štebih, D., Milavec, M., Gruden, K. and Žel, J. 2013. Quantitative analysis of food and feed samples with droplet digital PCR. *PLoS ONE* 8(5): e62583.
Mullis, K. B. 1990. The unusual origin of the polymerase chain reaction. *Sci. Am.* 262(4): 56–61, 64.
Nagamine, K., Hase, T. and Notomi, T. 2002. Accelerated reaction by loop-mediated isothermal amplification using loop primers. *Mol. Cell. Probes* 16(3): 223–229.
Niedzwiecka, A., Boucharef, L., Hahn, S., Zarske, M., Steinhilber, A., Poetz, O., Zagon, J., Seidler, T., Braeuning, A. and Lampen, A. 2018. A novel antibody-based enrichment and mass spectrometry approach for the detection of species-specific blood peptides in feed matrices. *Food Control* 98: 141–149.
Niessen, L. 2018. The application of loop-mediated isothermal amplification (LAMP) assays for the rapid diagnosis of food-borne mycotoxigenic fungi. *Curr. Opin. Food Sci.* 23: 11–22.

Norris, K. H., Barnes, R. F., Moore, J. E. and Shenk, J. S. 1976. Predicting forage quality by infrared reflectance spectroscopy. *J. Anim. Sci.* 43(4): 889–897.

Nyrén, P., Pettersson, B. and Uhlén, M. 1993. Solid phase DNA minisequencing by an enzymatic luminometric inorganic pyrophosphate detection assay. *Anal. Biochem.* 208(1): 171–175.

Ocaña, M. F., Neubert, H., Przyborowska, A., Parker, R., Bramley, P., Halket, J. and Patel, R. 2004. BSE control: detection of gelatine-derived peptides in animal feed by mass spectrometry. *Analyst* 129(2): 111–115.

Okorski, A., Polak-Śliwińska, M., Karpiesiuk, K., Pszczółkowska, A. and Kozera, W. 2017. Real time PCR: a good tool to estimate mycotoxin contamination in pig diets. *World Mycotoxin J.* 10(3): 219–228.

Omar, J., Boix, A. and von Holst, C. 2015. Differentiation of coccidiostats-containing feed additives by mid and near infrared microscopy. *Food Addit. Contam. A* 32(9): 1464–1474.

Omar, J., Slowikowski, B., Boix, A. and von Holst, C. 2017. Spectroscopy applied to feed additives of the European Union Reference Laboratory: a valuable tool for traceability. *Food Addit. Contam. A* 34(8): 1272–1284.

Ortea, I., O'Connor, G. and Maquet, A. 2016. Review on proteomics for food authentication. *J. Proteomics* 147: 212–225.

Oseland, E., Bish, M., Spinka, C. and Bradley, K. 2020. Examination of commercially-available bird feed for weed seed contaminants. *Invasive Plant Sci. Manag.* 13(1): 14–22.

Ottoboni, M. and Pinotti, L. 2017. Light microscopy in combination with computer image analysis for the identification of processed animal protein in feed. In: *Microscopy and Imaging Science: Practical Approaches to Applied Research and Education*, 746p A. Méndez-Vilas, (Ed.). Formatex Research Center S.L. ISBN: 9788494213496.

Ouyang, Z. and Cooks, R. G. 2009. Miniature mass spectrometers. *Annual Rev. Anal. Chem.* 2(1): 187–214.

Owens, B., McCann, M. E. E., Park, R. and McCracken, K. J. 2007. Defining feed wheat quality for broilers. *HGCA Res. Rep.* 423, 1–116.

Pacocha, N., Scheler, O., Nowak, M. M., Derzsi, L., Cichy, J. and Garstecki, P. 2019. Direct droplet digital PCR (dddPCR) for species specific, accurate and precise quantification of bacteria in mixed samples. *Anal. Methods* 11(44): 5730–5735.

Paracchini, V., Petrillo, M., Reiting, R., Angers-Loustau, A., Wahler, D., Stolz, A., Schonig, B., Matthies, A., Bendiek, J., Meinel, D. M., Pecoraro, S., Busch, U., Patak, A., Kreysa, J. and Grohmann, L. 2017. Molecular characterization of an unauthorized genetically modified Bacillus subtilis production strain identified in a vitamin B2 feed additive. *Food Chem.* 230: 681–689.

Parasoglou, P., Parrott, E., Zeitler, J., Rasburn, J., Powell, H., Gladden, L. and Johns, M. 2009. Quantitative moisture content detection in food wafers. In: *October 34th International Conference on Infrared, Millimeter, and Terahertz Waves* (vols. 21–25), Busan and Korea 3–4. doi: 10.1109/ICIMW.2009.5324623.

Park, R. S., Agnew, R. E., Gordon, F. J. and Steen, R. W. J. 1998. The use of near infrared reflectance spectroscopy (NIRS) on undried samples of grass silage to predict chemical composition and digestibility parameters. *Anim. Feed Sci. Technol.* 72(1-2): 155–167.

Pearson, T. C. and Wicklow, D. T. 2006. Detection of corn kernels infected by fungi. *Trans. ASABE* 49(4): 1235–1245.

Pinotti L., Ottoboni M., Caprarulo V., Giromini C., Gottardo D., Cheli F., Fearn T. and Baldi A. 2015. Microscopy in combination with image analysis for characterization of fishmeal material in aquafeed. *Animal Feed Science and Technology* 215: 156–164.

Piras, C., Roncada, P., Rodrigues, P. M., Bonizzi, L. and Soggiu, A. 2016. Proteomics in food: quality, safety, microbes and allergens. *Proteomics* 16(5): 799–815.

Piskata, Z., Servusova, E., Babak, V., Nesvadbova, M. and Borilova, G. 2019. The quality of DNA isolated from processed food and feed via different extraction procedures. *Molecules* 24(6): 1188.

Planque, M., Arnould, T. and Gillard, N. 2017. Food allergen analysis: detection, quantification and validation by mass spectrometry. In: *Allergen* Athari, S. S. (Ed). Intechopen. doi: 10.5772/intechopen.69361.

Qin, J., Xie, L. and Ying, Y. 2015. Determination of tetracycline hydrochloride by terahertz spectroscopy with PLSR model. *Food Chem.* 170: 415–422.

Qudoos, A., Bayram, I. and Iqbal, A. 2019. Determination of adulteration in animals feed by using FT-NIR Technology. In: Proceedings of the at the III International Eurasian Agriculture and Natural Sciences Congress, October 17–20, 2019, Antalya, Turkey. ISBN: 9786056901010.

Raes, M., Planque, M., Marbaix, H., Gillard, N., Lecrenier, M. C., Dieu, M., Mauro, S., Arnould, T. and Renard, P. 2018. Mass spectrometry based Approaches for Analysing Proteins in Processed Food and Feed. *Ann. Nutr. Food Sci.* 2: 1028.

Rasinger, J. D., Marbaix, H., Dieu, M., Fumière, O., Mauro, S., Palmblad, M., Raes, M. and Berntssen, M. H. G. 2016. Species and tissues specific differentiation of processed animal proteins in aquafeeds using proteomics tools. *J. Proteomics* 147: 125–131.

Rech, J. 2011. *Microscopie des plantes consommées par les animaux*. Versailles, France: Quae éditions. ISBN: 9782759209255.

Reece, P., Chassaigne, H., Collins, M., Buckley, M., Bremer, M. and Grundy, H. H. 2012. Proteomic analysis of meat and bone meal and animal feed. Namur: presses universitaires de Namur. In: *Detection, Identification and Quantification of Processed Animal Proteins in Feedingstuffs* Jørgensen, J. S. and Baeten, V. (Eds), 113–124. Namur: Presses universitaires de Namur. ISBN: 9782875510297.

Russ, J. C. 2005. *Image Analysis of Food Microstructure*. Boca Raton, FL, CRC Press Press. ISBN: 9781845691516.

Sanchez-Pinto, L. N., Luo, Y. and Churpek, M. M. 2018. Big data and data science in critical care. *Chest* 154(5) : 1239–1248.

Sanger, F., Nicklen, S. and Coulson, A. R. 1977. DNA sequencing with chain-terminating inhibitors. *Proc. Natl Acad. Sci. U. S. A.* 74(12): 5463–5467.

Santos, C. A., Lopo, M., Páscoa, R. N. and Lopes, J. A. 2013. A review on the applications of portable near-infrared spectrometers in the agro-food industry. *Appl. Spectrosc.* 67(11): 1215–1233.

Sattler, V. A., Mohnl, M. and Klose, V. 2014. Development of a strain-specific real-time PCR assay for enumeration of a probiotic *Lactobacillus reuteri* in chicken feed and intestine. *PLoS ONE* 9(2): e90208.

Scauflaire, J., Godet, M., Gourgue, M., Liénard, C. and Munaut, F. 2012. A multiplex real-time PCR method using hybridization probes for the detection and the quantification of *Fusarium proliferatum*, *F. subglutinans*, *F. temperatum*, and *F. verticillioides*. *Fungal Biol.* 116(10): 1073–1080.

Selvaraj, V., Maheshwari, Y., Hajeri, S. and Yokomi, R. 2019. Droplet digital PCR for absolute quantification of plant pathogens. In: *Plant Biotechnology: Progress in*

Genomic Era, 583–595 Khurana, S.M.P and Gaur, R. K. (Eds). Singapore: Springer. ISBN: 139789811384981.

Sentandreu, M. A. and Sentandreu, E. 2011. Peptide biomarkers as a way to determine meat authenticity. *Meat Sci.* 89(3): 280–285.

Seo, S.-H. and Cho, S.-J. 2016. Changes in allergenic and antinutritional protein profiles of soybean meal during solid-state fermentation with Bacillus subtilis. *LWT* 70: 208–212.

Shang, Y., Xu, W., Wang, Y., Xu, Y. and Huang, K. 2017. Novel multiplex qualitative detection using universal primer-multiplex-PCR combined with Pyrosequencing. *Food Chem.* 237: 773–778.

Shan, L., Haseeb, H. A., Zhang, J., Zhang, D., Jeffers, D. P., Dai, X. and Guo, W. 2019. A loop-mediated isothermal amplification (LAMP) assay for the rapid detection of toxigenic Fusarium temperatum in maize stalks and kernels. *Int. J. Food Microbiol.* 291: 72–78.

Shen, G., Fernández Pierna, J. A., Baeten, V., Cao, Y. and Yang, Z. 2020. Local anomaly detection and quantitative analysis of contaminants in soybean meal using near infrared imaging: the example of non-protein nitrogen. *Spectrochim. Acta A* 225: 117494-117509.

Shehata, H. R., Ragupathy, S., Shanmughanandhan, D., Kesanakurti, P., Ehlinger, T. M. and Newmaster, S. G. 2019. Guidelines for validation of qualitative real-time PCR methods for molecular diagnostic identification of probiotics. *J. AOAC Int.* 102(6): 1774–1778.

Shenge, K. C., Adhikari, B. N., Akande, A., Callicott, K. A., Atehnkeng, J., Ortega-Beltran, A., Kumar, P. L., Bandyopadhyay, R. and Cotty, P. J. 2019. Monitoring Aspergillus flavus genotypes in a multi-genotype aflatoxin biocontrol product With quantitative Pyrosequencing. *Front. Microbiol.* 10: 2529.

Sierra, V., Aldai, N., Castro, P., Osoro, K., Coto-Montes, A., Oliván, M. 2008. Prediction of the fatty acid composition of beef by near infrared transmittance spectroscopy. *Meat Sci.* 78(3): 248–255.

Sinnaeve, G., Dardenne, P., Agneessens, R. and Biston, R. 1994. The use of near infrared spectroscopy for the analysis of fresh grass silage. *J.Near Infrared Spectrosc.* 2(2): 79–84.

Song, T., Yang, L., Qiu, H. and Qiu, Z. 2017. Loop-mediated isothermal amplification assay for identification of animal derived materials. *J. Food Saf. Qual.* 8(8): 3207–3212.

Speranskaya, A. S., Krinitsina, A. A., Shipulin, G. A., Khafizov, K. F. and Logacheva, M. D. 2018. High-throughput sequencing for the authentication of food products: problems and perspectives. *Russ. J. Genet.* 54(9): 1003–1012.

Staats, M., Arulandhu, A. J., Gravendeel, B., Holst-Jensen, A., Scholtens, I., Peelen, T., Prins, T. W. and Kok, E. 2016. Advances in DNA metabarcoding for food and wildlife forensic species identification. *Anal. Bioanal. Chem.* 408(17): 4615–4630.

Steinhilber, A. E., Schmidt, F. F., Naboulsi, W., Planatscher, H., Niedzwiecka, A., Zagon, J., Braeuning, A., Lampen, A., Joos, T. O. and Poetz, O. 2018. Mass spectrometry-based immunoassay for the quantification of banned ruminant processed animal proteins in vegetal feeds. *Anal. Chem.* 90(6): 4135–4143.

Steinhilber, A. E., Schmidt, F. F., Naboulsi, W., Planatscher, H., Niedzwiecka, A., Zagon, J., Braeuning, A., Lampen, A., Joos, T. O. and Poetz, O. 2019. Application of mass spectrometry-based immunoassays for the species- and tissue-specific quantification of banned processed animal proteins in feeds. *Anal. Chem.* 91(6): 3902–3911.

Streit, E., Schatzmayr, G., Tassis, P., Tzika, E., Marin, D., Taranu, I., Tabuc, C., Nicolau, A., Aprodu, I., Puel, O. and Oswald, I. P. 2012. Current situation of mycotoxin contamination and co-occurrence in animal feed–focus on Europe. *Toxins* 4(10): 788–809.

Tähkäpää, S., Maijala, R., Korkeala, H. and Nevas, M. 2015. Patterns of food frauds and adulterations reported in the EU rapid alert system for food and feed and in Finland. *Food Control* 47: 175–184.

Tan, Z., Lu, J., Luo, J. and Xie, J. 2014. Continuous-wave terahertz imaging applied to detect infestations caused by insects in grain. *Adv. J. Food Sci. Technol.* 6(2): 271–274.

Tao, F., Yao, H., Hruska, Z., Burger, L. W., Rajasekaran, K. and Bhatnagar, D. 2018. Recent development of optical methods in rapid and non-destructive detection of aflatoxin and fungal contamination in agricultural products. *Trends Anal. Chem.* 100: 65–81.

Tena, N., Fernández Pierna, J. A., Boix, A., Baeten, V. and von Holst, C. 2014. Differentiation of meat; bone meal from fishmeal by near-infrared microscopy: extension of scope to defatted samples. *Food Control* 43: 155–162.

Theuer, M., Harsha, S. S., Molter, D., Torosyan, G., Beigang, R. 2011. Terahertz Time-Domain spectroscopy of gases, liquids, and solids. *ChemPhysChem* 12(15): 2695–2705.

Tran, H., Salgado, P., Tillard, E., Dardenne, P., Nguyen, X. T. and Lecomte, P. 2010. "Global" and "local" predictions of dairy diet nutritional quality using near infrared reflectance spectroscopy. *J. Dairy Sci.* 93(10): 4961–4975.

Ung, B., Fischer, B. M., Ng, B. W. H. and Abbott, D. 2009. Comparative investigation of detection of melamine in food powders. In 34th International Conference on Infrared, Millimeter, and Terahertz Waves, October 21–25, 2009, Busan, Korea: 1–2. doi: 10.1109/ICIMW.2009.5324952.

van Raamsdonk, L. W. D., Prins, T. W., Meijer, N., Scholtens, I. M. J., Bremer, M. G. E. G. and de Jong, J. 2019. Bridging legal requirements and analytical methods: a review of monitoring opportunities of animal proteins in feed. *Food Addit. Contam.* 36(1): 46–73.

Vermeulen, P., Ebene, M. B., Orlando, B., Fernández Pierna, J. A. and Baeten, V. 2017. Online detection and quantification of particles of ergot bodies in cereal flour using near-infrared hyperspectral imaging. *Food Addit.Contam. A* 29(2): 232–240.

Vermeulen, P., Fernández Pierna, J. A., van Egmond, H. P., Zegers, J., Dardenne, P. and Baeten, V. 2013. Validation; transferability study of a method based on near-infrared hyperspectral imaging for the detection; quantification of ergot bodies in cereals. *Anal. Bioanal. Chem.* 405(24): 7765–7772.

Vermeulen, P. J. A., Pierna, F., van Egmond, H. P., Dardenne, P., and Baeten, V. 2012. Online detection; quantification of ergot bodies in cereals using near infrared hyperspectral imaging. *Food Addit. Contam.* 29(2): 232–240.

Veys, P. and Baeten, V. 2010. New approach for the quantification of processed animal proteins in feed using light microscopy. *Food Addit. Contam.* 27(7): 926–934.

Veys, P. and Baeten, V. 2018. Protocol for the isolation of processed animal proteins from insects in feed and their identification by microscopy. *Food Control* 92: 496–504.

Veys, P., Berben, G. and Baeten, V. 2011. EURL-AP interlaboratory study microscopy 2010–Validation of a revised version of Annex VI of EU Regulation EC/152/2009 and proficiency evaluation. Final version. Gembloux, Belgium: CRA-W. Available at: http://www.eurl.craw.eu/img/page/interlaboratory/EURL_AP_ILS_2010_final.pdf (Last accessed 17/04/2020).

Von Bargen, C., Dojahn, J., Waidelich, D., Humpf, H. U. and Brockmeyer, J. 2013. New sensitive high-performance liquid chromatography-tandem mass spectrometry method for the detection of horse and pork in halal beef. *J. Agric. Food Chem.* 61(49): 11986–11994.

Wang, W. and Paliwal, J. 2005. Potential of near-infrared hyperspectral reflectance imaging for screening of farm feed contamination. *Proc. SPIE.* 5969: 644–653.

Willems, S., Fraiture, M. A., Deforce, D., De Keersmaecker, S. C., De Loose, M., Ruttink, T., Herman, P., Van Nieuwerburgh, F. and Roosens, N. 2016. Statistical framework for detection of genetically modified organisms based on next generation sequencing. *Food Chem.* 192: 788–798.

Williams, P. J., Geladi, P., Britz, T. J. and Manley, M. 2012. Investigation of fungal development in maize kernels using NIR hyperspectral imaging and multivariate data analysis. *J. Cereal Sci.* 55(3): 272–278.

Yang, Z., Han, L., Fernández Pierna, J. A. F., Dardenne, P. and Baeten, V. 2011. The potential of near infrared microscopy to detect, identify and quantify processed animal by-products. *J. Near Infrared Spectrosc.* 19(4): 211–231.

Yang, Z., Han, L., Wang, C., Li, J., Fernández Pierna, J. A., Dardenne, P. and Baeten, V. 2016a. Detection of melamine in soybean meal using near-infrared microscopy imaging with pure component spectra as the evaluation criteria. *J. Spectrosc.* 2016: 1–11.

Yang, Z., Wang, C., Han, L., Li, J. and Liu, X. 2014. Rapid screening and visual tracing of melamine in soybean meal by NIR microscopy imaging. *J. Innov. Opt. Heal. Sci.* 7(4): 1350072:1350082.

Yang, C., Zhang, J., Zhou, C., Huo, L., Cui, H. and Wang, L. 2016b. Research progress on the application of digital PCR in detection of foodborne pathogenic microorganisms. *J. Food Saf. Qual.* 7(12): 4726–4730.

Yang, Q., Domesle, K. J. and Ge, B. 2018. Loop-mediated isothermal amplification for Salmonella detection in food and feed: current applications and future directions. *Foodborne Pathog. Dis.* 15(6): 309–331.

Yates, J. R., Ruse, C. I. and Nakorchevsky, A. 2009. Proteomics by mass spectrometry: approaches, advances, and applications. *Annual Review of Biomedical Engineering* 11: 49–79.

Zhan, H., Xi, J., Zhao, K., Bao, R. and Xiao, L. 2016. A spectral-mathematical strategy for the identification of edible and swill-cooked dirty oils using terahertz spectroscopy. *Food Control* 67: 114–118.

Zhang, J., Wang, J., Cui, J. and Chen, X. 2018. Detection of transgenic rapeseed line Ms8 by loop-mediated isothermal amplification. *J. Food Saf. Qual.* 9(21): 5592–5595.

Zhao, L., Fang, P., Tang, J., Ma, Y., Li, Z., Wang, C., Su, Q., Jia, J. T. and Jiang, Y. 2017. Detection of Listeria monocytogenes in foods by droplet digital PCR. *J. Food Saf. Qual.* 8(11): 4133–4138.

Zou, M. Q., Zhang, X. F., Qi, X. H., Ma, H. L., Dong, Y., Liu, C. W., Guo, X. and Wang, H. 2009. Rapid authentication of olive oil adulteration by Raman spectrometry. *J. Agric. Food Chem.* 57(14): 6001–6006.

Part 2

Quality and safety assurance

Chapter 5

Developments in techniques to test the efficacy of animal feed products

Gerhard Flachowsky and Ulrich Meyer, Institute of Animal Nutrition - Friedrich-Loeffler-Institut (FLI), Germany

1 Introduction

Feed, water and feed additives may all be considered animal feed products. They are the most important inputs and cost factors in animal nutrition. There have long been rules to assess safety and efficacy of feeds and feed additives. As early as 1906, the US Congress passed the Pure Food and Drug Act, facilitating the development of safety regulations for food and feed. The Food and Drug Administration (FDA) was founded to develop and enforce these regulations as the primary federal regulatory agency in the United States responsible for the safety of food of animal origin. It manages this responsibility within the Animal Feed Safety System (AFSS). Similar developments can be observed in other countries. As an example, rules about the safety of food of animal origin were discussed in Germany at the end of the 19th century, although the first feed law was not adopted by the German Parliament until 1926. Since then the scope of regulation has been extended from safety to take in regulations relating to the efficacy of feed products.

A major development in Europe was the bovine spongiform encephalopathy (BSE) crisis (associated 'mad cow disease') from 1996 to 1998.

http://dx.doi.org/10.19103/AS.2020.0083.11

In 2002, the European Food Safety Authority (EFSA), based in Brussels (Belgium) was established in order to separate risk assessment from risk management. In 2005, the EFSA moved to Parma (Italy). The EFSA is currently divided into 10 panels. The 'Additives and Products or Substances used in Animal Feed (FEEDAP) Panel' provides scientific advice on the safety and/or the efficacy of additives and products or substances used in animal feed. The panel evaluates safety for users, consumers, animals and the environment and also assesses efficacy concerning the target animal species.

In addition to the FEEDAP-Panel, there are other EFSA-Panels connected to animal feed products, animal health and welfare as well as safety of food of animal origin for humans and environment, such as:

- 'Panel on Animal Health and Welfare (AHAW-Panel)'; this panel provides scientific advice on all aspects of animal diseases and welfare, mainly for food producing animals;
- 'Panel on Contaminants in the Food Chain (CONTAM Panel)'; this panel provides scientific advice on contaminants in the food chain and undesirable substances such as natural toxicants, mycotoxins and residues of unauthorised substances;
- 'Panel on Nutrition, Novel Foods and Food Allergens (NDA-Panel)'; this panel deals with questions related to human nutrition, novel foods, nutrient sources and food for special groups;
- 'Panel for Food Additives and Flavourings (FAF-Panel)'; this panel evaluates the safety of chemical substances added to food and consumer exposure to these substances.

Where appropriate, the FEEDAP-Panel liaises with one or more of the other panels in order to guarantee the safety of certain additives for humans as workers or consumers, animals and the environment.

All feed additives for the European market (as well as those derived from outside of Europe) can only be authorised after their safety and efficacy have been assessed by the EFSA. All the procedures for evaluation and authorisation of feed additives are laid down in the regulation (EC) No 1831/2003 on additives for use in animal nutrition and its implementing rules (Regulation (EC) No 429/2008). Studies/experiments by researchers/applicants should be performed according to EFSA-Guidance documents. Every five years the FEEDAP-Panel reviews and adopts each of its rules for the assessment of the safety and efficacy of feed additives in animal nutrition. The last adoption was conducted in 2018 and this version (EFSA, 2018) replaces the previous one. All applicants have to submit their applications to the EU Commissions according to the latest guidance document. All studies for the development of any animal feed additive must currently be performed according to this EFSA-Guidance document (EFSA, 2018).

The objective of this chapter is to describe developments relating to testing the efficacy of animal feed products. The chapter describes the rules prescribed by the EFSA and other organisations (e.g. the AFSS of the FDA) governing studies to test safety and efficacy of animal feed products. The chapter includes examples of studies for protein/amino acid supply, mycotoxin decontamination and methane inhibition.

2 The EFSA-FEEDAP Guidance document

Animal feed products are subdivided into five main groups (EFSA, 2018):

1 Technological additives;
2 Sensory additives;
3 Nutritional additives;
4 Zootechnical additives; and
5 Coccidiostats and histomonostats.

These groups are summarised below.

Most additives belong to the technological additives group, including:

- Preservatives;
- Antioxidants;
- Emulsifiers;
- Stabilisers;
- Thickeners;
- Gelling agents;
- Binders;
- Anti-caking agents;
- Acidity regulators;
- Silage additives;
- Denaturants; and
- Hygiene condition enhancers.

In the EFSA-Guidance document (EFSA, 2018) the FEEDAP-Panel distinguishes between technological additives which exert their function in feed, such as silage additives, preservatives, antioxidants and stabilisers, and substances which exert their effect in the animal, such as substances for the reduction of contamination of feed by mycotoxins or substances for reduction of radionuclides.

The group of sensory additives includes:

- Colour added to feedingstuffs specifically for colouring of animal products (such as eggs);

- Colour added to feedingstuffs;
- Substances to colour ornamental fish and birds; and
- Flavoring compounds.

Nutritional additives, such as amino acids and their salts, urea, vitamins and provitamins as well as trace minerals, belong to the group of essential nutrients. Animal nutritional requirements/needs are determined by scientific bodies, for example the Society of Nutrition Physiology (GfE) or the National Research Council (NRC) of the USA (GfE, 1995, 1999, 2001, 2008, 2014, NRC, 1994, 2001, 2007, 2012).

Zootechnical additives cover substances which may influence animal yield, feed efficacy or product quality, such as:

- Additives affecting animal production, such as enzymes (e.g. phytases, polysaccharidases and proteases);
- Additives favourably affecting the environmental consequences of animal production (such as additives that reduce N or P excretion and methane emissions);
- Additives affecting the characteristics of food of animal origin; and
- Additives affecting animal welfare.

Coccidiostats and histomonostats do not belong to the group of feed additives. They protect animals from the consequences of infection by *Eimeria spp*. or *Histomonas meleagridis* and are given in feed (Holdsworth et al., 2004, Chapman, 1998). They are therefore also included in the group of feed additives.

EFSA (2018) recommends *in vivo* efficacy studies for the most important animal species/categories and includes guidelines for acceptable studies in physiologically-related species (see Table 1). Study results should be extrapolated to physiologically related species in order to reduce the number of animals used in experiments (EMA, 2017). Animal feeding studies should be carried out according to the 3R-principle: Replacement, Reduction, Refinement (DFG, 2016, EMA, 2017, GfE, 2017). Before animal experimentation, laboratory, *in vitro* and/or *in sacco* studies should be undertaken in order to account for the 3R-principle and to reduce animal numbers in such experiments. Results should also be extrapolated between physiologically similar animal species and age groups. There should be similarities in gastrointestinal function and in metabolism.

For efficacy studies with feed additives the EFSA (2018) recommends at least three feeding studies using the 'main species' and investigation of an adequate numbers of animals. Studies should demonstrate statistically significant results for key parameters. More details are given for the minimum duration of long-term efficacy studies (see Table 2).

Table 1 Extrapolation of efficacy data from certain (so-called main) species to other physiologically related species (EFSA, 2018)

'Main species'	Extrapolated to physiologically related species
Dairy cows	Other dairy ruminants (e.g. goats, sheep, buffalos)
Calves or cattle for fattening	Other growing ruminants (e.g. sheep, goats, buffalos) at the corresponding developmental stage
Sows	Other reproductive *Suidae*
Piglets or pigs for fattening	Other growing *Suidae*
Chickens for fattening (broilers)	Other poultry for fattening (e.g. turkeys, ducks, geese, pheasants, quails, guinea fowls, ostriches) and ornamental birds
Laying hens	Other birds kept for egg production or breeding (e.g. turkeys, ducks, geese, pheasants, quails, guinea fowls, ostriches)
Salmon or trout	Ornamental fish
Horses	Other Equidae
Rabbits	Other Leporidae

Establishing the minimum duration of experiments (see Table 2) always involves a compromise between scientific requirements, on the one side, and financial budgets, available animal experimental facilities, 3-R -principles etc., on the other side. The duration values listed in Table 2 are therefore minimum values. For dairy cows, for example, the whole lactation period including the dry period would be much more informative than the 84 days mentioned in the guidelines (see Table 2). For minor species not included in Table 2, the duration of the studies should correspond to that of the physiologically related major species listed earlier. In some cases, longer experimental periods could be useful to effectively study the animals' period of adaptation to the additive(s) (e.g. in the case of methane inhibitors in ruminants; see Section 4.3).

The endpoints of studies depend on the effects that are expected from the additives. For all studies, the following parameters should be measured:

- Feed intake (total dry matter, concentrate, roughage in the case of ruminants);
- Initial and final body weight;
- Body weight gain (slaughtering results);
- Milk yield and laying performance (in the case of lactating and laying animals);
- Feed to gain (yield) ratio;
- Water intake in case the additive is administered via water; and

Table 2 Proposals for minimum duration of long-term efficacy studies (EFSA, 2018)

Animal species/ category	Definition of the animal category	Start	Minimum duration
Dairy cows	Lactating cows	4-8 weeks after calving	84 days
Calves	Reared for reproduction, veal or beef production	1-4 weeks of age	56 days
Cattle	Bovine animals, completed the weaning period	Full rumination, but <6 months of age	84 days
Lambs/kids	Young animals reared for reproduction or meat	1-4 weeks of age	56 days
Sheep/goats	Lactating animals	4 weeks of parturition	84 days
Piglets (weaned)	Young animals having completed the suckling period	<7 days after weaning	42 days (35 days, if growth rate >0.5 kg/day)
Pigs for fattening	Animals for meat production until slaughtering	<35 kg	Until slaughter, but not less than 70 days
Sows	Female animals having been inseminated	Insemination/Mating	For effects on reproduction: two cycles
Chickens for fattening	Birds raised for fattening	1 day of age	35 days
Laying hens	Female birds held for egg production	22-25 weeks	84 days
Turkeys for fattening	Birds raised for fattening	1 day old	84 days
Salmon/trouts	Growing salmonids	Trout: >19 g Salmon: >50 g	84 days
Rabbits	Reared for reproduction or meat production	1 week after birth	42 days
Breeding does	At least one pregnancy	Insemination (mating)	Two cycles for effects on reproduction
Cats, dogs or other non-food-producing animals			28 days

- Clinical observations including general health status, morbidity and mortality should also be monitored.

Depending on the animal species and the category, the parameters listed in Table 3 should also be measured during the whole experimental period.

Product quality and the composition of milk, meat and eggs should be determined and compared with the product quality of control animals which have not received the additive. Important parameters are:

- Composition of the animal product (including possible residues of undesirable substances);
- Physical/technological properties (e.g. water-binding capacity, oxidative stability);
- Sensory modification of food product (e.g. colour, taste, smell, texture or other subjective criteria assessed by a trained panel); and
- Hygiene quality of food (e.g. numbers of spoilage organisms, potential human or animal enteropathogens).

Measures of faecal consistency, mainly dry matter content, can also be used as an objective measurement. The composition of the faeces (dry matter, N, P, K etc.) can be determined as well. Environmental effects, such as reduction of methane or ammonia emission, or N, P or S excretion, are also objective parameters to measure various excretions.

In addition to the *in vivo* studies as described above, there are also some other types of study that can be used in order to assess the effects of feed additives in animals, such as:

- Literature reviews; and
- *In vitro* studies.

Both types of studies may contribute to the analysis of the effects of the additive and to understand the mode of action, but they cannot replace animal feeding studies using the specific animal feed product. Normally, such studies should be carried out before experiments are undertaken.

3 Reporting of efficacy studies

According to the current 'Guidance on the assessment of the efficacy of feed additives' (EFSA, 2018) reporting/publication of efficacy studies should meet the following criteria:

1 Title, summary, objectives and materials and methods should be clearly described.

2 An ethical statement indicating compliance with national and institutional guidelines for care and use of animals should be given, following RRR principles (see DFG, 2016, EMA, 2017, GfE, 2017).
3 Animals, housing and husbandry. Requirements include the following:
  - Species, breed, age, initial body weight, sex, identification procedure, physiological stage and general health should be stated.
  - Husbandry conditions including feeding and rearing conditions (pen/tank size, stocking density, temperature, lighting) should be recorded and reported, where relevant; for aquatic species, conditions include water quality, including water flow rate, water temperature and salinity.
  - In terms of diets, there should be a description of the manufacture and quantitative composition of diet(s) in terms of ingredients used, relevant nutrients (analysed values) end energy (digestible, metabolisable or net energy). For studies with enzymes, diets should be analysed for enzyme content and the enzyme-specific substrates.
4 Study design. Study location (including dates and responsible individuals), study duration and type of study design (factorial, stratified, crossover) should be mentioned. The number of treatments and control groups should be indicated. Numbers of replicates (experimental unit) per group and number of animals per replicate should be provided. The guidelines suggest that the 3R-principle should be considered in the experimental design. The experimental unit (individual, pen etc.) and the basis for different measurements should be indicated for each parameter. The basis for the selection of the number of animals/replicates and power analysis should be explained. The intended concentration of the active substance(s) or agent(s) in the feed should be determined.

Table 3 Additional parameters to be measured in animal feeding studies (EFSA, 2018)

Animal species/category	Additional parameters
Dairy animals	Milk production (including fat-corrected milk), milk composition (total solids, protein, fat, lactose, urea), protein, fat and lactose yield and somatic cell counts
Sows	Number of piglets born, piglets born alive, litter weight at birth and at weaning, number of piglets weaned, weaning to oestrus interval
Laying hens	Laying rate, egg weight, feed to egg mass ratio, egg mass/hen per day
Breeding hens	Laying rate, fertility, hatchability and chick viability
Fish	Specific growth rate

5 Experimental procedure. The procedures carried out to the different experimental groups should be detailed. These should include the parameters/endpoints measured, indicating when and how they were measured, as well as information on the methods of analysis. Animal health should be monitored, and morbidity and mortality (including culling) recorded. The methodology to correct feed to gain ratio for mortality (including culling) should be reported.
6 Statistical methods. The methods used to perform statistical analysis should be stated, including those used to identify outliers, and handle missing data. If any relevant data points are excluded from the model (e.g. outliers), a justification should be given. Any methods used to assess whether the data met the assumptions of the statistical approach should be described.
7 Results. The results of the study should be presented for all endpoints considered in the experiment. The concentration of the active substance(s) or agent(s) in the feed to which the additive is added should be periodically analysed and reported. Tables should be used to summarise the results from all treatments. For all endpoints which are measured on individual animals in a pen, a summary parameter of the endpoint in the experimental unit should be used (e.g. mean for continuous measurements such as body weight, median and counts for quarterly measurements such as severity of an outcome or mortality). Summary parameters should always be adjusted for losses (mortality/culling). The distribution of losses within the treatment groups should be assessed to avoid the risk of introducing a bias.
8 Discussion and conclusions. The interpretation of the results should consider the objectives of the study and the hypotheses underpinning additive function, including the mode of action of the substance (e.g. Duin et al. (2016). The results should be compared with other relevant studies in the literature. The authors should comment on any limitations of the study including any potential sources of bias, limitations in experimental design and any degree of imprecision associated with the results. Adequate safety studies should be included in the efficacy study. The conclusions from the study should be drawn based on the objective of the study, the hypothesis it sets out to test and the outcome of the study. A trial protocol data sheet can complete all the documents submitted by the applicant. The report of the efficacy study (see above) is intended to provide guidance for scientists and/or applicants to submit a discussion paper for authorisation of a substance as a feed additive for use in animal nutrition. The following chapters demonstrate some examples of steps in the research process.

4 Key research steps in demonstrating the efficacy of animal feed products

This and the following sections describe key research steps in developing animal feed products and demonstrating their efficacy. Three substances or substance groups are considered as examples:

- Nitrogen (N) compounds include amino acids as feed additives for food producing animals and consequences of deficits and excess for both animals and for the environment.
- Mycotoxins as antinutritive/toxic substances for animals in feed and the search for potential means to inactivate/destroy these substances.
- Methane (CH_4) production by ruminants as co-product of rumen fermentation and possible ways to reduce methane emissions: methane has a high greenhouse gas potential (about 23 x of CO_2; IPCC (2006).

The simple model in Figure 1 demonstrates inputs and outputs of animals which shows where these substances fit within the overall nutrient cycle. As noted earlier, all studies should be performed according to 3R-principles.

Key objectives for these studies are as follows:

- The objective in the study of nitrogen (N) input is a maximum retention of N in the animal body or in food of animal origin and minimal excretion to the environment (see Section 5 and Tables 4 and 5).
- In the case of mycotoxins, the objective is for all these harmful substances to be destroyed or inactivated before feed is consumed by animals (see Section 6 and Tables 6 and 7).
- As far as methane is concerned, it has the highest greenhouse gas (GHG) potential as a fermentation product of microbial activity in the rumen, though it is also produced in smaller amounts in the large intestine. The objective is to reduce GHG-fermentation processes and thus damaging emissions to the environment (see Section 7 and Tables 8–10).

These are discussed in the following sections.

5 Demonstrating efficacy: the case of reducing nitrogen (N) excretion

One of the main objectives of animal husbandry is to produce edible protein for human nutrition. Table 4 shows animal species/categories and their daily yields of edible protein per kg body weight and day. The lowest protein yield per kg of body weight is in growing ruminants and pigs.

Large proportions of consumed nitrogen are excreted by ruminants and non-ruminants. The highest portion of consumed nitrogen (between 55 and 90%) is excreted via urine and faeces, as shown in Table 5. In ruminants excreted amounts vary between 65% und 90% of N-intake, depending on animal species, categories and the performance of individual animals (NRC, 1985, Flachowsky and Lebzien, 2006, Šebek et al., 2014, Velthof, 2014). Similar percentages have also been measured for non-ruminants.

Studies were undertaken to find possibilities to reduce N-excretion of ruminants and non-ruminants (see recommendations of GfE and requirements of NRC listed in the reference list).

There are some opportunities to reduce N-excretion by food-producing animals. These include:

- Meet the amino acids requirements of nonruminants (see GfE-recommendations and NRC-requirements listed in the reference list) and add adequate amounts of amino acids;
- Avoid N-excess in the rations of ruminants and meet the N-requirements of ruminal microorganisms; add rumen protected amino acids, if necessary;
- Higher animal performances may increase N-excretion per animal and day, but reduce it per unit food of animal origin (see Tables 4 and 5);
- Shorter growing periods for heifers and longer productive lifespan of cows reduce the N-excretion per kg milk;
- Understand the N-metabolism in the rumen and the amino acids metabolism including the splanchnic metabolism (gluconeogenesis) in high yielding cows and take into consideration this knowledge;
- Shorter growing periods for young sows and longer productive lifespan for sows

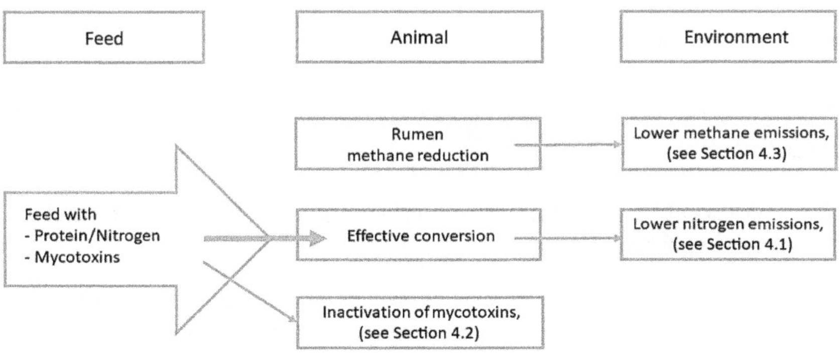

Figure 1 Effect of feed treatment and animal metabolism on reduction of N-excretion (see Section 4.1), reduction of contamination with mycotoxins (see Section 4.2) and reduction of methane emissions (see Section 4.3).

Table 4 Effects of animal species/categories and performances on yields of edible protein (Niemann et al., 2011)

Protein source (BW)	Yield per day	DM[a] -intake (kg/day)	Protein in edible fraction (g/kg)	Edible protein (g/day)	Edible protein (g/kg of BW)
Dairy cow (650 kg)	10kg milk	12	34	323	0.7
	20kg milk	16		646	1.0
	40kg milk	25		1292	2.0
Beef cattle (350 kg)	500g WG[b]	6.5	190	48	0.14
	1000g WG	7.0		95	0.27
	1500g WG	7.5		143	0.41
Growing/fattening pig (80 kg BW[c])	500g WG	1.8	150	45	0.56
	700g WG	2.0		63	0.80
	1000g WG	2.2		81	1.0
Broiler (1.5 kg BW)	40g WG	0.07	200	4.8	3.2
	60g WG	0.08		7.2	4.8
Laying hen (1.8 kg BW)	50% LP[d]	0.10	120	3.4	1.9
	70% LP	0.11		4.8	2.7
	90% LP	0.12		6.2	3.4

[a]Dry matter, [b]Weight gain in g/day, [c]Body weight, [d]Laying performance in %.

- N-/protein supply should be according to animal yield and overnutrition must be avoided. Amino acids and in ruminants rumen protected amino acids should be added to the diet.
- Don't add amino acids to the water for drinking, if the amino acid requirements are already met by feeding (Kamphues et al., 2019).

Presently there exist *in vivo, in sacco* and *in vitro* methods to assess protein degradation and microbial protein synthesis in the rumen. In the future, *in vivo* methods will be replaced increasingly by *in vitro/in sacco* methods calibrated based on results of *in vivo* studies (RRR-principles).

6 Demonstrating efficacy: the case of reducing mycotoxin contamination of feed

Mycotoxins are small and stable toxic secondary metabolites produced by fungal organisms. They are considered one of the most significant hazards in human and animal nutrition (EFSA, 2009, Santos Pereira et al., 2019, Ráduly et al., 2020). About one-fourth of cereals produced globally is assumed to be

Table 5 Effects of animal species, categories and daily yields on N-excretion (in urine and faeces; Niemann et al., 2011)

Animals species/categories	Performance per day	N-Excretion (% of intake)
Dairy cow (650kg)	10kg of milk	75
	20kg of milk	70
	40kg of milk	65
Beef cattle (350kg BW[a])	500g WG[b]	90
	1000g WG	84
	1500g WG	80
Growing/fattening pig (80kg BW)	500g WG	85
	700g WG	80
	1000g WG	75
Broiler (1.5kg BW)	40g WG	70
	60g WG	60
Laying hen (1.8kg BW)	50% LP[c]	80
	70% LP	65
	90% LP	55

[a]Body weight, [b]Weight gain in g/day, [c]Laying performance in %.

Table 6 Most relevant endpoints for substances reducing the contamination of feed by mycotoxins (EFSA, 2018)

Mycotoxin	Most relevant endpoint(s)
Alfatoxin B1	Aflatoxin in animal products (e.g. milk, egg yolk)
Deoxynivalenol (DON)	DON/metabolites in blood serum
Zearalenone	Zearalenone (+ α- and ß-zearalenone in plasma, excretion of zearalenone/metabolites)
Ochratoxin A	Ochratoxin in kidney (or blood serum)
Fumonisins B_1 + B_2	Sphinganine/sphingosine ratio in blood, plasma or tissues

contaminated with mycotoxins (Puri et al., 2019). They are capable of causing potentially fatal diseases in humans and animals. Important mycotoxins are aflatoxins, deoxynivalenol, fumonisine, ochratoxin A, zearalenone and Ergot alkaloids (EFSA, 2009, Kamle et al., 2019, Mahato et al., 2019). Since 2002, the EFSA has issued more than 40 scientific opinions on the risks of mycotoxins to human and animal health (Eskola et al., 2018). Some mould species are able to produce different mycotoxins. The mycotoxin content of feed depends on the type of feed and climatic conditions (Döll and Dänicke, 2011; Adeyeye, 2016; Santos Pereira et al., 2019).

Table 7 Steps to demonstrate the effects of substances for reduction of the contamination of feed by mycotoxins (Eskola et al., 2018)

Stage	Type of studies	Endpoints
1	*In vitro* studies	Studies should demonstrate the intended effect of the additive (e.g. effects on mycotoxin(s); effects in the gastrointestinal-tract of nonruminants; effects in the rumen (Rusitec)
2	*In vivo* studies under Institute conditions	At least three independent *in vivo* studies (generally short-term studies) should be performed in at least two different locations. The studies should show significant effects at the lowest recommended dose. Mycotoxin binder should be also given to the unsupplemented (free of mycotoxins) control animals.
3	*In vivo* studies under field conditions	For additives intended to be used in all terrestrial species, efficacy should be demonstrated *in vivo* in 3 major species, representing different digestive systems (e.g. poultry species, non-ruminant mammals and ruminants)

Čolović et al. (2019) have summarised the differing methods to reduce the contamination of food and feed:

- Cleaning products (mainly seeds);
- Thermal processing;
- Chemical agents;
- Feed additives to prevent the effects of mycotoxins; and
- Biological detoxification and biotransformation.

Other methods such as cold atmospheric plasma treatment also have the potential to reduce mycotoxin concentration in food and feed as demonstrated for aflatoxin B_1 and fumonisin B_1 (Hojnik et al., 2017, Wielogorska et al., 2019).

EFSA guidance suggests various ways to demonstrate the efficacy of substances in reducing mycotoxin contamination of animal feed (Eskola et al., 2018):

- Degradation/inactivation/transformation of mycotoxins;
- Reduction of the absorption of mycotoxins;
- Increased excretion of mycotoxins; and
- Reduced concentration of mycotoxins in food of animal origin.

Degradation and other forms of inactivation of mycotoxins are the main means of preventing contamination of animal feed (Dänicke et al., 2012, Lorenz et al., 2019, Kamle et al., 2019). As an alternative, mycotoxin binders or adsorbents are substances that bind to mycotoxins and prevent them from being absorbed through the gut and impede their entry into the blood stream (EFSA, 2009). These binders may be (Eskola et al., 2018):

- Organic or inorganic substances, such as activated carbon;
- Aluminium silicates, such as bentonite, montmorillonite and zeolite;
- Complex indigestible-carbohydrates; and
- Other synthetic polymers.

De Mil et al. (2015) for example tested 27 mycotoxin binders and their effect on *in vitro* zearalenone adsorption; Nazarizadeh and Pourreza (2019) evaluated three mycotoxin binders to prevent the adverse effects of aflatoxin B_1 in growing broilers.

Table 6 shows the most relevant endpoints for assessing efficacy in reducing mycotoxin contamination, whilst Table 7 shows steps to demonstrate efficacy. *In vivo* studies, Stage 2 in Table 7, should include the animal category for which the lowest maximum content of the respective mycotoxin in feed is proposed.

There is an additional point concerning the supplementation of mycotoxin-contaminated feed with mineral adsorbents. Adsorbents such as bentonite, zeolite, montmorillonite or hydrated Na-Ca-aluminum silicate can bind or adsorb mycotoxins to interlayer species or external surfaces. Several studies have indicated that these agents may also cause undesirable effects in farm animals (mainly pigs and poultry) as summarised by Elliott et al. (2020). These undesirable effects include binding of essential micronutrients and vitamins in feed, resulting in reduced feed conversion, lower productivity in livestock and possibly cytotoxic effects and immunosuppression. The main weakness of a number of studies regarding the efficacy of mineral binders is the missing mycotoxin-free diet supplemented with mineral binders (Döll and Dänicke, 2004). The following groups should therefore be included in Stage 2 feeding studies (see Table 7):

- Unsupplemented control group;
- Control group plus mycotoxin binder;
- Control group with mycotoxin; and
- Control group with mycotoxin and mycotoxin binder.

The most successful binder should be tested under field conditions (see Table 7).

7 Demonstrating efficacy: the case of reducing methane emissions from the rumen

The livestock sector is a significant source of greenhouse gases such as methane (CH_4). Methane is an unavoidable by-product of anaerobic microbial fermentation of carbohydrates in the rumen of ruminants (Hungate, 1967,

Johnson and Johnson, 1995, Hristov et al., 2013). It is produced in considerable amounts, representing between 4% and 10% of the gross energy consumed with the diet with extremes of 2% and 15% (Flachowsky and Brade, 2007, Tamminga et al., 2007). It contributes between 50% and 80% of the carbon footprint (CF) of food of ruminant origin because of its high greenhouse gas potential (about 23 x CO_2: IPCC, 2006).

Energy losses due to CH_4 losses from the rumen have been known for a long time (Tappeiner, 1882). Table 8 shows estimates of the impact of methane reduction on energy intake and milk yield in dairy cows. Over many decades animal nutritionists have tried to reduce methane losses from the rumen through strategies such as diet, increasing performance and livestock management (Tamminga et al., 2007; Bannink et al., 2008; Niemann et al., 2011; Patra et al., 2017). There has been growing political pressure to reduce methane emissions from ruminants to help tackle the challenge of climate change (Steinfeld et al., 2006).

In response to these challenges, animal nutritionists have tried to reduce methane emissions from the rumen of ruminants in various ways as summarized in Table 9. Table 9 shows three main approaches:

- High concentrate diets;
- Fats, fatty acids and oilseeds; and
- Feed additives.

In each case it suggests their potential to contribute to further reductions in emissions and where further research may be beneficial.

In the category of feed additives, some phytogenic substances and other chemicals could have the potential to reduce methane emissions from ruminants (Patra et al., 2017; Broucek, 2018). Ionophores also have some methane-reduction potential (e.g. Odongo et al., 2007; Thompson et al., 2019). Beauchemin and McGinn (2006) measured 9% reduction of methane emission after supplementation of monensin to beef cattle. In a meta-analysis of monensin effects in beef cattle, Appuhamy et al. (2013) found a reduction of methane emissions of 19 g per animal per day. In a more recent study, Williams et al. (2020) added 800 g cottonseed oil or 400 g tannin or both per day to diets of dairy cows and reduced methane yield by 14%, 11% and 20%, respectively. The supplementation of monensin as feed additive is not currently permitted in the EU. In the case of many potential additives, long-term methane reduction potential has not yet been well established (Hristov et al., 2013).

3-nitrooxy propanol (3-NOP) has recently attracted interest in methane reduction research. Duval and Kindermann (2011) concluded that 3-NOP inhibited methane formation by targeting the active site of methyl coenzyme-M reductase. Lopes et al. (2016) and others have tested the influence of 3-NOP on rumen fermentation and methane emission and found a decreased methane emission per kg dry matter intake of 34%. A meta-analysis (involving 11 experiments and 38 treatments) by Dijkstra et al. (2018) quantified the effects of 3-NOP on CH_4-production in the rumen of beef and dairy cattle. 10 mg 3-NOP/kg DM reduced CH_4 emissions by 38.8% compared to unsupplemented animals. Jayanegara et al. (2018) carried out 12 *in vivo* studies with dairy cows, beef cattle and sheep and found similar results. Van Wesemael et al. (2019) applied 1.6 g of 3-NOP/cow and day via roughage or concentrate and found 28 or 23% lower CH_4-production for roughage and concentrate, respectively. These promising studies highlight the potential for finding substances with effective methane reduction potential.

A five-stage programme can be used to assess the effects of substances on methanogenesis (see Table 10). Characterisation of the substance(s) starts with description of the plant(s) and the manufacturing process, followed by analytical methods, identification of the substance(s) and their functional groups, their purity, and a documented history of safe use (Stage 1; Table 10). The analytical characterisation of phytogenic substances is the next step in assessment (Stage 2; Table 10). Most phytogenic additives are a mixture of substances. Chemical analysis of substances should be the starting point of nutritional assessment of phytogenic feed additives. Researchers should try to characterise all substances with their chemical structures, their functional groups and their purity. In the case of mixture of substances, one or more key substances and their quantity in the mixture should be clearly described. For example, a description of essential oils as preparations containing mixtures of natural-identical essential oils is generally not scientifically acceptable. The proportion of essential oils in the preparation should be listed. All steps of processing of the phytogenic substances should be clearly described. In the case of chemicals used as methane inhibitors, there is no need for stages 1 and 2, but stages 3–5 should be carried out according to the recommendations outlined in this section.

The next step (Stage 3, Table 10) deals with the nutritional assessment of phytogenc material *in vitro*. *In vitro* systems should be used for screening a group of substances and they may also contribute to studying the mode of action. Advantages of screening are to assess a broad spectrum of phytogenic substances in numerous dosages. Disadvantages of screening include the elimination of substances without any *in vitro* effect, but possibly with CH_4 reduction potential *in vivo*. Further pros and cons of *in vitro* studies are discussed by Flachowsky and Lebzien (2012). Some phytogenic substances may contain ingredients which inhibit rumen methanogenesis but may also reduce feed intake and decrease cell wall digestibility. An example are halogenous derivates which inhibit methanogenesis *in vitro* (e.g. Trei and Scott, 1971), but were found to reduce feed intake and animal growth *in vivo* (Chalupa, 1977; Sawyer et al., 1971) and potentially alter the microbial community in the

Table 8 Model calculation to show the influence of methane reduction on the energy available for dairy cows and milk yields (conditions for calculation: dry matter intake (DMI): 20 kg/cow and day; body weight: 650 kg/cow; energy intake: 7 MJ NEL/kg DM with 20 g CH_4 emission; Niemann et al., 2011)

	Methane reduction			
	30g/kg DMI	25g/kg DMI	20g/kg DMI	15g/kg DMI
Item	600g/(cow/d)	500g/(cow/d)	400g/(cow/d)	300g/cow/d)
Energy intake (MJ/d)	130	135	140	145
Milk yield (kg/d)	28.0	29.5	31.0	32.5
CH_4-emission (g/kg)	21.4	17.0	12.9	9.2
Carbon footprint (g CO_2/kg milk)	735	585	440	315

Table 9 Specific measures/activities to reduce methane emissions from ruminants, significance for Europe and assessment of research needs (see also Flachowsky and Lebzien, 2012, Hristov et al., 2013)

Measures	Significance for Europe	Assessment of research need
Feeding of high concentrate diets (high starch, low fibre)	potential nearly exhausted	medium
Feeding of fats, fatty acids, oilseeds	potential nearly exhausted	low
Feed additives:		
Halogen compounds (e.g. Cl, Br)	not permitted in the EU	zero
Ionophores (e.g. monensin)	not permitted in the EU[a]	zero
H_2-binding substances (e.g. fumaric acid, acrylic acid)	presently no signifcance	low
Phytogenic substances or plants containing such substances (e.g. tannins, saponins, essential oils, herbs)	presently no signifcance	high
Further feed additives such as yeasts, enzymes	presently no signifcance	low
Further chemicals (e.g. 3-nitrooxypropanol (3-NOP)	presently not permitted	high

[a]Presently only permitted as veterinary drug for dairy cows in the EU.

rumen (Martinez-Fernandez et al., 2016; Martinez-Fernandez et al., 2018). *In vitro* studies should therefore be considered as an initial point for screening of potential methane inhibitors and must be combined with *in vivo* experiments (Stages 4 and 5).

Depending on the results of *in vitro* studies, a potentially effective phytogenic substance should be examined *in vivo*, if possible in several dosages to investigate effects on feed intake, digestibility, animal health, methane emission and animal performances. As already mentioned, many studies have demonstrated a high variability of results between *in vitro* (Stage 3) and *in vivo* (Stage 4) studies (Flachowsky and Lebzien, 2012). The relationship between methane produced *in vivo* and *in vitro* in our own studies ($r^2 = 0.264$; Flachowsky and Lebzien, 2009) was very poor and further measurements are needed. Studies to measure feed intake, *in vivo* digestibility and methane emissions are important endpoints of Stage 4 (Table 10).

It is sometimes difficult to distinguish between plant effects based on their chemical composition and effects caused by phytogenic components. Archimède et al. (2011) conducted a meta-analysis of a database of 112 treatments to compare effects of C3 and C4 grasses as well as legumes grown in warm or cold climate on CH_4-production in ruminants. Results indicate that ruminants fed C4-grasses produced 17% more methane/kg dry matter compared to ruminants fed C3-grasses. Animals fed legumes grown in tropical regions produced 20% less methane than those fed with legumes of cold regions. In legumes from warm regions highly lignified cell walls and the presence of secondary metabolites such as tannins decreased cell wall digestion and their nutritive value as well as CH_4-production. A clear description of the plants used in the basal diet and an assessment of the content of specific substances with potential influence on methanogenesis are therefore necessary.

If phytogenic substances are effective in the target animal species/categories (Stage 4, Table 10), then they should be fed to target species (e.g. differing species or breeds of ruminants) in long-term feeding experiments (Stage 5, Table 10). Such studies are very important for nutritional assessment of the additive(s) and for reliable recommendations to the animal feed industry and to the farmers. Various feed rates of the additive should be tested. Apart from investigating animal yield, long-term dose-response studies in target animals are also necessary to assess the influence of phytogenic substance(s) on animal health, fertility and composition/quality of food of animal origin. Energy balances are potentially useful in order to demonstrate the efficient use of feed for animal yield rather than methane production.

Table 10 Proposal of stages for nutritional and safety assessment of phytogenic feed additives (Flachowsky and Lebzien, 2012)

Stage	Type of studies	Endpoints
1	Characterisation of plant(s)	Description of plant(s), cultivation conditions, vegetation stage of harvest, processing etc.
2	Analytical characterisation	Chemical structure of the molecule(s) and their functional group(s), characterisation of impurities, analytical determination in premixes and diets
3	*In vitro* studies	Testing of the effects of additives on rumen fermentation under various conditions (dose-level studies, various rations etc.), screening of various substances, possible deduction of mode of actions
4	*In vivo* studies	Influence of additive(s) on feed intake, rumen fermentation, digestibility, methane emissions etc. mostly within rumen-fistulated animals. Dose-response studies are recommended
5	Long-term feeding studies with target animal species/categories	Influence of additive(s) on animal health, yield and fertility; assessment of adaption of rumen microbes to the additive(s), safety and quality of food of animal origin, environmental impact, reproduction of results

8 Conclusions

Studies to test the efficacy of animal feed products are critical to demonstrate their effectiveness. Some examples are shown in this chapter. These studies should be carried out according to current EFSA guidance documents (EFSA, 2018) in order for results to support potential applications of new feed additives.

9 References

Adeyeye, S. A. O. 2016. Fungal mycotoxins in foods: a review. *Cogent Food and Agriculture* 2(1), 1213127.

Appuhamy, J. A. D. R. N., Strathe, A. B., Jayasundara, S., Wagner-Riddle, C., Dijkstra, J., France, J. and Kebreab, E. 2013. Anti-methanogenic effects of monensin in dairy and beef cattle: a meta-analysis. *Journal of Dairy Science* 96(8), 5161–5173.

Archimède, H., Eugène, M., Marie Magdeleine, C., Boval, M., Martin, C., Morgavi, D. P., Lecomte, P. and Doreau, M. 2011. Comparison of methane production between C3 and C4 grasses and legumes. *Animal Feed Science and Technology* 166–167, 59–64.

Bannink, A., France, J., Lopez, S., Gerrits, W. J. J., Kebreab, E., Tamminga, S. and Dijkstra, J. 2008. Modelling the implications of feeding strategy on rumen fermentation and functioning of the rumen wall. *Animal Feed Science and Technology* 143(1–4), 3–26.

Beauchemin, K. A. and McGinn, S. M. 2006. Effects of various feed additives on the methane emissions from beef cattle. *International Congress Series* 1293, 152–155.

Broucek, J. 2018. Options to methane production abatement in ruminants: a review. *Journal of Animal and Plant Sciences* 28, 348–364.

Chalupa, W. 1977. Manipulating rumen fermentation. *Journal of Animal Science* 45(3), 585–599.

Chapman, H. D. 1998. Evaluation of the efficacy of anticoccidial drugs against Eimeria species in the fowl. *International Journal for Parasitology* 28(7), 1141–1144.

Čolović, R., Puvača, N., Cheli, F., Avantaggiato, G., Greco, D., Đuragić, O., Kos, J. and Pinotti, L. 2019. Decontamination of mycotoxin-contaminated feedstuffs and compound feed. *Toxins* 11(11), 617.

Dänicke, S., Kersten, S., Valenta, H. and Breves, G. 2012. Inactivation of deoxynivalenol-contaminated cereal grains with sodium metabisulfite: a review of procedures and toxicological aspects. *Mycotoxin Research* 28(4), 199–218.

De Mil, T., Devreese, M., De Baere, S., Van Ranst, E., Eeckhout, M., De Backer, P. and Croubels, S. 2015. Characterization of 27 mycotoxin binders and the relation with in vitro zearalenone adsorption at a single concentration. *Toxins* 7(1), 21–33.

DFG 2016. *Animal Experimentation in Research*, Bonn, DFG Publ., 76 p. ISBN 978-3-527-33570-1.

Dijkstra, J., Bannink, A., France, J., Kebreab, E. and Van Gastelen, S. 2018. Short communication: antimethanogenic effects of 3-nitrooxypropanol depend on supplementation dose, dietary fiber content, and cattle type. *Journal of Dairy Science* 101(10), 9041–9047.

Döll, S. and Dänicke, S. 2004. In vivo Detoxification of Fusarium toxins. *Archives of Animal Nutrition* 58(6), 419–441.

Döll, S. and Dänicke, S. 2011. The Fusarium toxins deoxynivalenol (DON) and zearalenone (ZON) in animal feeding. *Preventive Veterinary Medicine* 102(2), 132–145.

Duin, E. C., Wagner, T., Shima, S., Prakash, D., Cronin, B., Yanez-Ruiz, D. R., Duval, S., Rumbeli, R., Stemmler, R. T., Thauer, R. K. and Kindermann, M. 2016. Mode of action uncovered for the specific reduction of methane emissions from ruminants by the small molecule 3-nitrooxypropanol. *Proceedings of the National Academy of Sciences of the United States of America* 113(22), 6172–6177.

Duval, S. and Kindermann, M. 2011. Use of nitrooxy organic molecules in feed for reducing methane emission in ruminants, and/or to improve ruminant performance. International Publication Number WO 2012/084629 Al. European Patent Appl. No 111 78994.7.

EFSA 2009. Supporting publication: review of mycotoxin-detoxifying agents used as feed additives: mode of action, efficacy and feed. *Food Safety* 6, 22E.

EFSA 2018. Guidance on the assessment of the efficacy of feed additives. *EFSA Journal. European Food Safety Authority* 16(5), e05274.

Elliott, C. T., Connolly, L. and Kolawole, O. 2020. Potential adverse effects on animal health and performance caused by the addition of mineral adsorbents to feeds to reduce mycotoxin exposure. *Mycotoxin Research* 36(1), 115–126.

EMA 2017. *Guideline on the Principles of Regulatory Acceptance of 3Rs (Replacement, Reduction, Refinement) Testing Approaches*, European Medicines Agency. Available at https://www.ema.europa.eu/en/regulatory-acceptance-3r-replacement-reduction-refinement-testing-approaches. Accessed on 06.01.2021.

Eskola, M., Altieri, A. and Galobart, J. 2018. Overview of the activities of the European Food Safety Authority on mycotoxins in food and feed. *World Mycotoxin Journal* 11(2), 277–289.

Flachowsky, G. and Brade, W. 2007. Potenziale zur Reduzierung der Methan-Emissionen bei Wiederkäuern. *Züchtungskunde* 79, 417–465.

Flachowsky, G. and Lebzien, P. 2006. Possibilities for reduction of nitrogen (N) excretion from ruminants and the need for further research–a review. *Landbauforschung Völkenrode* 56, 19–30.

Flachowsky, G. and Lebzien, P. 2009. Comments on in vitro studies with methane inhibitors. *Animal Feed Science and Technology* 151(3–4), 337–339.

Flachowsky, G. and Lebzien, P. 2012. Effects of phytogenic substances on rumen fermentation and methane emissions: a proposal for a research process. *Animal Feed Science and Technology* 176(1–4), 70–77.

GfE 1995. *Empfehlungen zur Energie und Nährstoffversorgung der Mastrinder*, Frankfurt am Main, DLG-Verlag.

GfE 1999. *Empfehlungen zur Energie und Nährstoffversorgung der Legehennen und Masthühner (Broiler)*, Frankfurt am Main, DLG-Verlag.

GfE 2001. *Empfehlungen zur Energie und Nährstoffversorgung der Milchkühe und Aufzuchtrinder*, Frankfurt am Main, DLG-Verlag.

GfE 2008. *Recommendations for the Supply of Energy and Nutrients to Pigs*, Frankfurt am Main, DLG-Verlag.

GfE 2014. *Recommendations for the Supply of Energy and Nutrients to Horses*, Frankfurt, DLG-Verlag.

GfE 2017. Stellungnahme zur Unerlässlichkeit von Tierversuchen und zur Eignung von Ersatzmethoden in der Tierernährungsforschung. *Proceedings of the Society of Nutrition Physiology*, 218-224.

Hojnik, N., Cvelbar, U., Tavčar-Kalcher, G., Walsh, J. L. and Križaj, I. 2017. Mycotoxin decontamination of food: cold atmospheric pressure plasma versus "classic" decontamination. *Toxins* 9(5), 151.

Holdsworth, P. A., Conway, D. P., McKenzie, M. E., Dayton, A. D., Chapman, H. D., Mathis, G. F., Skinner, J. T., Mundt, H. C., Williams, R. B. and World Association for the Advancement of Veterinary Parasitology. 2004. World Association for the Advancement of Veterinary Parasitology (WAAVP) guidelines for evaluating the efficacy of anticoccidial drugs in chickens and turkeys. *Veterinary Parasitology* 121(3-4), 189-212.

Hristov, A. N., Oh, J., Firkins, J. L., Dijkstra, J., Kebreab, E., Waghorn, G., Makkar, H. P. S., Adesogan, A. T., Yang, W., Lee, C., Gerber, P. J., Henderson, B. and Tricarico, J. M. 2013. Mitigation of methane and nitrous oxide emissions from animal operations: a review of enteric methane mitigation options. *Journal of Animal Science* 91(11), 5045-5069.

Hungate, R. E. 1967. Hydrogen as an intermediate in the rumen fermentation. *Archiv für Mikrobiologie* 59(1), 158-164.

IPCC. 2006. *Guidelines for National Greenhouse Gas Inventories: Agriculture, Forestry and Other Land Use*, Hayama, Japan, INSTITUTE for global environmental strategies.

Jayanegara, A., Sarwono, K. A., Kondo, M., Matsui, H., Ridla, M., Laconi, E. B. and Nahrowi 2018. Use of 3-nitrooxypropanol as feed additive for mitigating enteric methane emissions from ruminants: a meta-analysis. *Italian Journal of Animal Science* 17(3), 650-656.

Johnson, K. A. and Johnson, D. E. 1995. Methane emissions from cattle. *Journal of Animal Science* 73(8), 2483-2492.

Kamle, M., Mahato, D. K., Devi, S., Lee, K. E., Kang, S. G. and Kumar, P. 2019. Fumonisins: impact on agriculture, food, and human health and their management strategies. *Toxins* 11(6), 328.

Kamphues, J., Flachowsky, G., Rieger, H. and Meyer, U. 2019. Für und Wider eine Verabreichung von Futtermittelzusatzstoffen über das Tränkwasser. *Übersichten zur Tierernährung* 43, 205-248.

Lopes, J. C., De Matos, L. F., Harper, M. T., Giallongo, F., Oh, J., Gruen, D., Ono, S., Kindermann, M., Duval, S. and Hristov, A. N. 2016. Effect of 3-nitrooxypropanol on methane and hydrogen emissions, methane isotopic signature, and ruminal fermentation in dairy cows. *Journal of Dairy Science* 99(7), 5335-5344.

Lorenz, N., Dänicke, S., Edler, L., Gottschalk, C., Lassek, E., Marko, D., Rychlik, M. and Mally, A. 2019. A critical evaluation of health risk assessment of modified mycotoxins with a special focus on zearalenone. *Mycotoxin Research* 35(1), 27-46.

Mahato, D. K., Lee, K. E., Kamle, M., Devi, S., Dewangan, K. N., Kumar, P. and Kang, S. G. 2019. Aflatoxins in food and feed: an overview on prevalence, detection and control strategies. *Frontiers in Microbiology* 10, 2266-2266.

Martinez-Fernandez, G., Denman, S. E., Yang, C., Cheung, J., Mitsumori, M. and McSweeney, C. S. 2016. Methane inhibition alters the microbial community, hydrogen flow, and fermentation response in the rumen of cattle. *Frontiers in Microbiology* 7, 1122.

Martinez-Fernandez, G., Duval, S., Kindermann, M., Schirra, H. J., Denman, S. E. and McSweeney, C. S. 2018. 3-NOP vs. halogenated compound: methane production, ruminal fermentation and microbial community response in forage fed cattle. *Frontiers in Microbiology* 9, 1-3.

Nazarizadeh, H. and Pourreza, J. 2019. Evaluation of three mycotoxin binders to prevent the adverse effects of aflatoxin b1 in growing broilers. *Journal of Applied Animal Research* 47(1), 135–139.

Niemann, H., Kuhla, B. and Flachowsky, G. 2011. Perspectives for feed-efficient animal production. *Journal of Animal Science* 89(12), 4344–4363.

NRC. 1985. *Ruminant Nitrogen Usage*, Washington, DC, The National Academies Press.

NRC. 1994. *Nutrient Requirements of Poultry* (9th rev. edn.), Washington DC, National Academic Press.

NRC. 2001. *Nutrient Requirements of Dairy Cattle* (7th rev. edn.), Washington, DC, National Academies Press.

NRC. 2007. *Nutrient Requirements of Small Ruminants: Sheep, Goats, Cervids, and New World Camelids*, Washington DC, National Academies Press.

NRC. 2012. *Nutrient Requirements of Swine. 11th Revised Editions*, Washington DC, National Academic Press.

Odongo, N. E., Bagg, R., Vessie, G., Dick, P., Or-Rashid, M. M., Hook, S. E., Gray, J. T., Kebreab, E., France, J. and Mcbride, B. W. 2007. Long-term effects of feeding monensin on methane production in lactating dairy cows. *Journal of Dairy Science* 90(4), 1781–1788.

Patra, A., Park, T., Kim, M. and Yu, Z. 2017. Rumen methanogens and mitigation of methane emission by anti-methanogenic compounds and substances. *Journal of Animal Science and Biotechnology* 8, 13.

Puri, S., Shingh, S. and Tiwari, P. 2019. Mycotoxins: a threat to food security and health. *International Journal of Applied Sciences and Biotechnology* 7(3), 298–303.

Ráduly, Z., Szabó, L., Madar, A., Pócsi, I. and Csernoch, L. 2020. Toxicological and medical aspects of aspergillus-derived mycotoxins entering the feed and food chain. *Frontiers in Microbiology* 10, 2908.

Santos Pereira, C., Cunha, S. C. and Fernandes, J. O. 2019. Prevalent mycotoxins in animal feed: occurrence and analytical methods. *Toxins* 11(5), 290.

Sawyer, M. S., Hoover, W. H. and Sniffen, C. J. 1971. Effects of a methane inhibitor on growth and energy metabolism in sheep–Abstracts of papers to be presented at the sixty-sixth annual meeting of the american dairy science association Michigan State University East Lansing June 20–23, 1971. *Journal of Dairy Science* 5, 792–793.

Šebek, L. B., Bikker, P., Van Vuuren, A. M. and Van Krimpen, M. 2014. *Nitrogen and Phosphorus Excretion Factors of Livestock - Task 2: In-Depth Analyses of Selected Country Reports*, Wageningen, The Netherlands, Wageningen University and Research Center Livestock Research.

Steinfeld, H., Gerber, P. J., Wassenaar, T., Castel, V., Rosales, M. and De Haan, C. 2006. *Livestock's Long Shadow: Environmental Issues and Options*, Roma, FAO, 380 p.

Tamminga, S., Bannink, A., Dijkstra, Z. and Zom, R. 2007. Feeding strategies to reduce methane losses in cattle. *Animal Science Group, Wageningen UR, Rep. 34*, 44 p.

Tappeiner, H. 1882. Vergleichende Untersuchungen der Darmgase. *Biological Chemistry* 6, 4–5.

Thompson, L. R., Beck, M. R., Gunter, S. A., Williams, G. D., Place, S. E. and Reuter, R. R. 2019. An energy and monensin supplement reduces methane emission intensity of stocker cattle grazing winter wheat. *Applied Animal Science* 35(4), 433–440.

Trei, J. E. and Scott, G. C. 1971. Performance of steers on the methane inhibitor HCS - Abstracts of Papers for Presentation at the 63rd Annual Meeting of the American Society of Animal Science. *Journal of Animal Science* 33, 301–302.

Van Wesemael, D., Vandaele, L., Ampe, B., Cattrysse, H., Duval, S., Kindermann, M., Fievez, V., De Campeneere, S. and Peiren, N. 2019. Reducing enteric methane emissions from dairy cattle: two ways to supplement 3-nitrooxypropanol. *Journal of Dairy Science* 102(2), 1780–1787.
Velthof, G. L. 2014. Report Task 1 of Methodological Studies in the Field of Agro-Environmental Indicators. Lot 1 Excretion Factors. *Alterra Report*, Wageningen, The Netherlands.
Wielogorska, E., Ahmed, Y., Meneely, J., Graham, W. G., Elliott, C. T. and Gilmore, B. F. 2019. A holistic study to understand the detoxification of mycotoxins in maize and impact on its molecular integrity using cold atmospheric plasma treatment. *Food Chemistry* 301, 125281.
Williams, S. R. O., Hannah, M. C., Eckard, R. J., Wales, W. J. and Moate, P. J. 2020. Supplementing the diet of dairy cows with fat or tannin reduces methane yield, and additively when fed in combination. *Animal* 10, s464–s472.

Chapter 6

Advances in understanding key contamination risks in animal feed: mycotoxins

Luciano Pinotti, Luca Ferrari, Nicoletta Rovere, Francesca Fumagalli, Sharon Mazzoleni and Federica Cheli, University of Milan, Italy

1 Introduction

Animal and human health are of major concern in the feed/food industries. Crops grown as feed ingredients must be safe in order to be placed on the market. Plant and/or fungal products such as mycotoxins are one of the major contaminants of natural origin in feed and food ingredients (Čolović et al., 2019; Gruber-Dorninger et al., 2019). Mycotoxins are toxic chemical compounds typically produced by fungi or mold (Bennett and Klich, 2003; Sun et al., 2017) (Fig. 1). These secondary metabolites are produced following sudden changes in temperature (e.g. between day and night), mechanical breakage of the mold itself, or because of an increase or decrease in humidity. Mycotoxins are a serious risk to human and animal health (Khaneghah et al., 2019). According to the FDA, the potential costs of crop losses due to mycotoxin contamination are an average of $932 million per year (Smith et al., 2016).

http://dx.doi.org/10.19103/AS.2020.0083.12

Figure 1 Fungal contamination of maize.

2 Mycotoxin groups

Mycotoxins are produced by several species of filamentous fungi, above all *Aspergillus*, *Penicillium* and *Fusarium* (Liew and Mohd-Redzwan, 2018). More than 400 mycotoxins are currently known (Liu and Applegate, 2020). Many of these toxins are now classified as 'emerging' as there is still insufficient evidence regarding their toxicity. Others are already widely known and, as a result, there are legal limits on the levels of contamination that cannot be exceeded (Cheli et al., 2013, 2014). From an agricultural point of view, there are six main mycotoxin groups (Gruber-Dorninger et al., 2019; Pinotti et al., 2016):

- aflatoxins,
- zearalenone (ZEN),
- type A trichothecenes (e.g. T-2 toxin),
- type B trichothecenes (e.g. deoxynivalenol [DON]),
- ochratoxins (e.g. ochratoxin A [OTA]), and
- fumonisins.

Aflatoxins, zearalenone, trichothecenes, ochratoxins and fumonisins are discussed in the following sections.

2.1 *Aflatoxins*

Aflatoxins are the most widespread mycotoxin group comprising 18 different compounds. There are five major toxins called B1, B2, G1, G2 and M1 (Bordin et al., 2014) (Fig. 2): the abbreviations B and G are due to the blue or green fluorescence that they emit when irradiated by a UV radiation of 360 nm. The letter M identifies the aflatoxin typically found in milk. Of these, aflatoxin B1 is the most potent natural carcinogen and mutagen which has been classified as a Group 1 human carcinogen by the International Agency for Research on Cancer (IARC) (Mupunga et al., 2017).

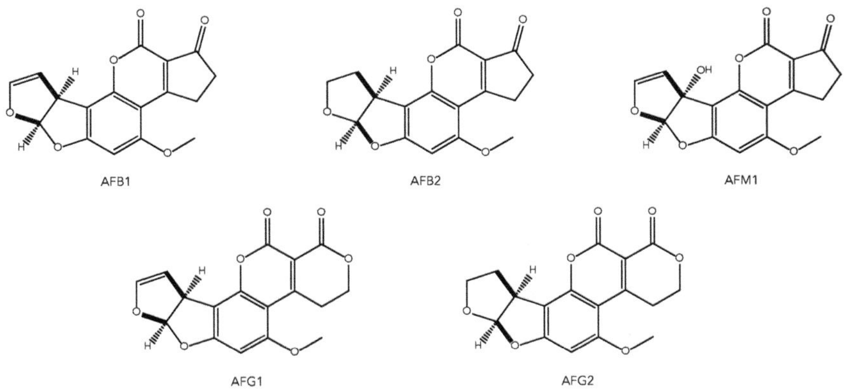

Figure 2 Chemical structure of aflatoxins.

The production of these toxins is the result of various factors, such as the fungal species, environment and substrate (Kew, 2013). These toxins are produced by fungi belonging to the *Aspergillus* genus at a temperature of between 25°C and 28°C. The production of toxins stops when the temperature drops below 13°C or rises above 42°C. The production of aflatoxins is particularly high in seasons and regions with high temperatures and low rainfall. Sub-Saharan Africa, for example, has a very high incidence, with 50% of raw cereal grains infected by Aflatoxin B1 (AFB1), while Northern Europe has an incidence of 6% (Kew, 2013). There is particular concern regarding the indirect exposure to aflatoxins through the consumption of contaminated products such as milk, eggs and meat (Marchese et al., 2018).

2.2 Zearalenone (ZEN)

Zearalenone (ZEN) is a lactone nonsteroidal estrogenic mycotoxin produced by several *Fusarium* and *Gibberella* species, above all *F. graminearum* (Iheshiulor et al., 2011; Lee and Ryu, 2017). Its structure is very similar to the female sex hormone and thus it is classified as a nonsteroidal hormone. ZEN is classified as a Group 3 carcinogen by the IARC due to its strong estrogenic activity (Alshannaq and Yu, 2017). ZEN is a white, crystalline and fat-soluble compound with a relatively high melting point (164–165°C) (Liu and Applegate, 2020).

The main exposure route to ZEN for humans and animals is through the consumption of cereal grains and cereal products contaminated by toxigenic fungal species, either contaminated directly in the field or during processing and storage (Zheng et al., 2018). Once absorbed, ZEN can be converted into different metabolites such as ZAN and zearalenol (Fig. 3). ZAN is synthetized by the removal of a double bond, while zearalenol is obtained through the removal of a ketonic group (EFSA, 2017a).

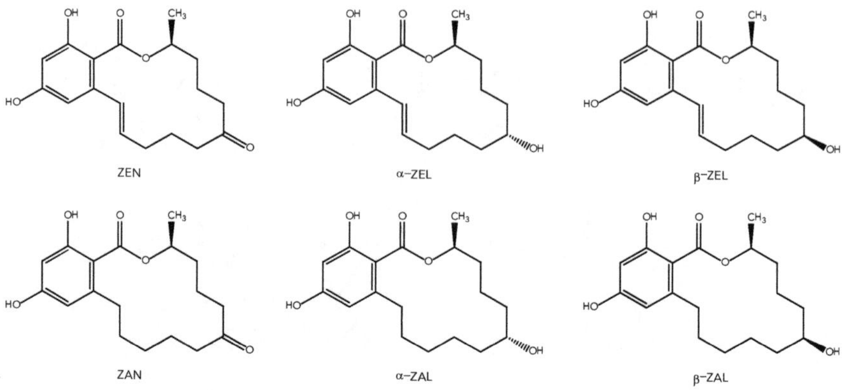

Figure 3 Zearalenone and its metabolites.

2.3 Trichothecenes

Some *Fusarium* spp., such as *F. culmorum* and *F. graminearum,* cause Fusarium Head Blight (FHB) – a destructive disease of cereal grain crops with a global economic impact (Parry et al., 1995). Trichothecenes are sesquiterpenoid mycotoxins that have been implicated in this disease. More than 150 compounds are classified into four types from A to D. Types A and B trichothecenes are the most important. Most toxic type A trichothecenes are HT-2 and T-2 toxins (Fig. 4) although, as it is not very prevalent, T-2 toxin is classified by IARC as Group 3 (WHO, 1993). Type B trichothecenes are frequently represented by DON (Fig. 5) and its derivatives. The International Agency for Research on Cancer has classified DON as a Group 3 carcinogenesis substance (Ostry et al., 2017).

In general, trichothecenes are easily absorbed via the integumentary and gastrointestinal system, and their ingestion can cause intestinal irritation in mammals which can lead to feed refusal in livestock. The T-2 toxin is roughly ten

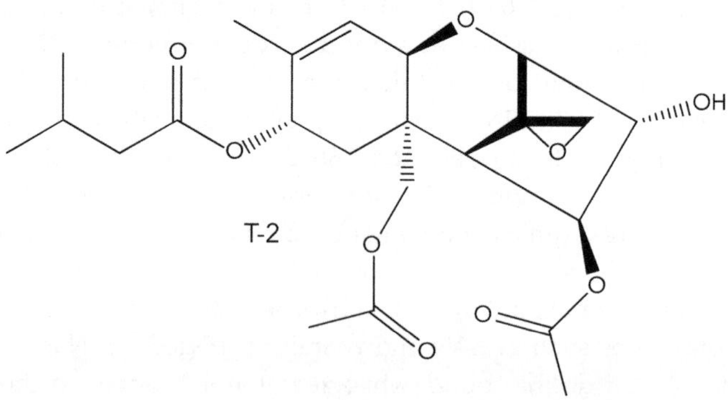

Figure 4 T-2 toxin structure.

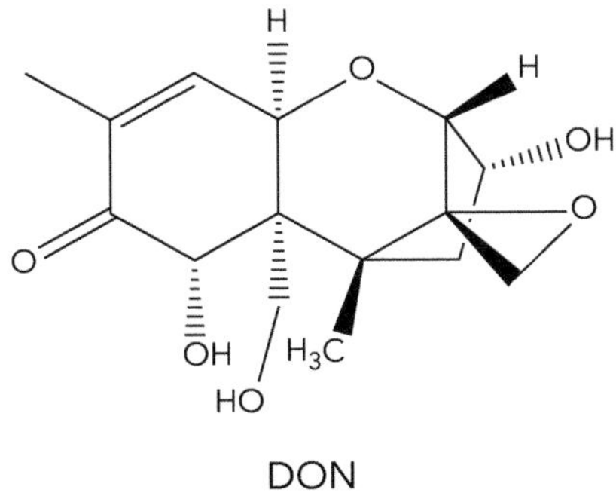

Figure 5 Chemical structure of deoxynivalenol.

times more toxic in mammals than DON (Foroud and Eudes, 2009). DON is the most prevalent toxin associated with FHB and belongs to the more phytotoxic type B trichothecenes (Langevin et al., 2004).

2.4 Ochratoxins

Several types of ochratoxins are produced by species of the genera *Aspergillus* and *Penicillium* (Denli and Perez, 2010), notably OTA, ochratoxin B (dechlorinated OTA) and ochratoxin C (ethylated OTA). Although they are often co-produced, the best known in this category are OTA and B. The difference between the two forms is the presence of a chlorine atom on the aromatic ring (Fig. 6).

OTA is the most important metabolite in terms of toxicology (Iheshiulor et al., 2011; Reddy and Bhoola, 2010). It is classified as a Group 2B potential human carcinogen by the International Agency for Research on Cancer (Reddy and Bhoola, 2010). Its carcinogenic action is probably the result of biotransformation due to cytochrome P450, which occurs both in humans and animals. Its absorption takes place in the gastrointestinal tract and, once in the blood, it binds to albumin (Kőszegi and Poór, 2016) with long residence times in the body.

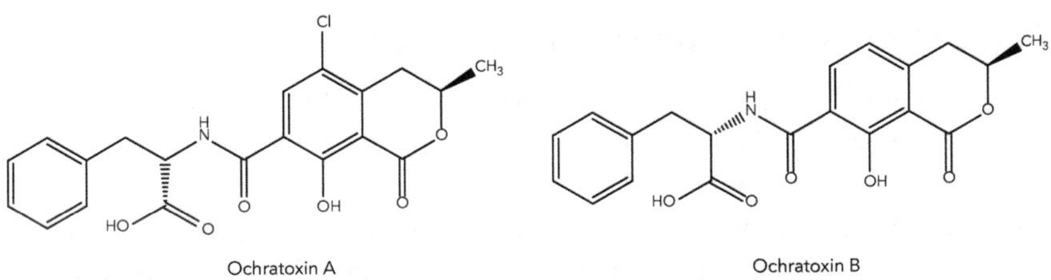

Figure 6 Ochratoxin A (left) and ochratoxin B (right).

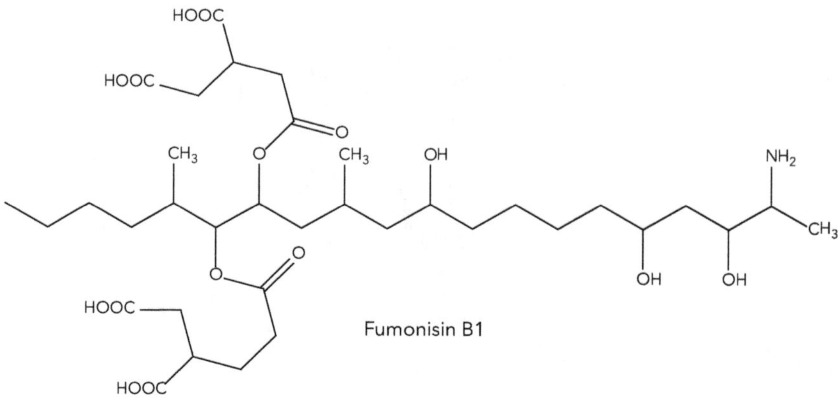

Figure 7 Fumonisin B1 structure.

2.5 Fumonisins

Fumonisins are a group of toxic substances mainly produced by fungi of the genus *Fusarium* in cereal products (Deepa and Sreenivasa, 2017), although *Aspergillus niger* also produces these toxins in cereals. As is the case with other toxins, fumonisins have high thermal resistance and are only destroyed at temperatures above 220°C.

More than 15 fumonisin homologues have been discovered and classified into groups A to P (Kamle et al., 2019). The molecular structure of fumonisins is well established and these toxins are very similar to sphingosine. The most important group is the B series, which includes B1 (Fig. 7), B2 and B3 toxins (Bertero et al., 2018). Of these, B1 is the most dangerous, which is classified in Group 2B (probably carcinogenic) by the IARC (Alshannaq and Yu, 2017), while fumonisins B2 and B3 are closely related metabolites and often coexist with their original structure (FB1). Other fumonisins have been recently discovered such as the FBX series and the 28 isomers of fumonisin B1. Toxins belonging to the FBX series consist of 12 compounds with the tricarballylic acid moiety replaced by other carboxylic acids, such as *cis*-aconitic acid, oxalylsuccinic acid and oxalylfumaric acid (Scott, 2012).

3 Regulation of mycotoxin levels in feed

The Joint Expert Committee on Food Additives (JECFA) – a scientific advisory body of the World Health Organization (WHO) and the FAO – has established regulatory threshold values for food and feed to protect consumers' health (Cheli et al., 2013; Čolović et al., 2019; Smith et al., 2016). Every country has fixed regulatory limits. In Europe, for example, aflatoxin B1 is regulated by the Commission Directive (EC) 100/2003 on undesirable substances in animal feed. For other mycotoxins, such as deoxynivalenol, ZEN, OTA and fumonisin B1 and B2, only non-binding recommendation values are determined for feeds in the Commission Recommendation 2006/576/EC (Table 1).

In the United States, the Federal Drug Administration (FDA) also recommends action levels for mycotoxins in animal feeds (Table 2). These recommendations are applied to aflatoxins, fumonisins and DON. Action levels have not been established for OTA and ZEN. Differences in acceptable mycotoxin levels between countries and regions can lead to distortions in international commodity markets, with more contaminated crops excluded from countries with stricter regimes and, potentially, traded more extensively in countries with less regulatory control (Pinotti et al., 2016; Piotrowska et al., 2013).

Table 1 European legislation on mycotoxins for cereals and cereal-based foods and feeds

	Maximum levels (ppb)					
Matrix	AFB1	OTA	DON	ZEN	F	Regulation
Cereals and cereal by-products (excluding maize by-products)	-	250	8000	2000	-	Commission Recommendation (EC) 576/2006
Maize by-products	-	-	12 000	3000	-	
Complete and complementary feedstuffs for pigs	-	50	900	250	-	
Complete and complementary feedstuffs for calves and lambs	-	-	2000	500	-	
Complete and complementary feedstuffs for poultry	-	100	-	-	-	
All feed materials	20	-	-	-	-	Commission Directive (EC) 100/2003
Complete feedstuffs for dairy animals	5	-	-	-	-	
Complete feedstuffs for calves and lambs	10	-	-	-	-	
Complete feedstuffs for pigs, poultry, cattle, sheep and goats	20	-	-	-	-	

-, limit not established; AFB1, aflatoxin B1; OTA, ochratoxin A; ZEN, zearalenone; DON, deoxynivalenol; F, fumonisins.

Table 2 The FDA's recommended action levels for aflatoxins in animal feeds and feed ingredients

Matrix	Maximum levels (ppb)			Regulation
	AF	DON	F	
Corn and peanut products for finishing beef cattle	300	-	-	Compliance Policy Guide 683.100
Cottonseed meal for beef cattle, swine and poultry	300	-	-	
Corn and peanut products for swine over 100 lb	200	-	-	
Corn and peanut products for breeding swine and mature poultry	100	-	-	
Animal feeds and ingredients (cottonseed meal excluded) for immature animals	20	-	-	
Animal feeds and ingredients for dairy animals, animals not listed above	20	-	-	
Corn and corn by-products not to exceed 20% of the diet for equids and rabbits	-	-	5000	
Corn and corn by-products not to exceed 50% of the diet for swine	-	-	20 000	
Corn and corn by-products not to exceed 50% of the diet for breeding ruminants, breeding poultry and breeding mink	-	-	30 000	
Corn and corn by-products not to exceed 50% of the diet for ruminants ≥3 months old raised for slaughter and mink raised for pelt production	-	-	60 000	
Corn and corn by-products not to exceed 50% of the diet for poultry raised for slaughter	-	-	100 000	
Corn and corn by-products not to exceed 50% of the diet for all other species or classes of livestock and pet animals	-	-	10 000	
Grain and grain by-products for ruminating beef and feedlot cattle older than 4 months	-	10 000 in feed ingredients and in the total ration	-	
Distillers grains, brewers grains, gluten meals, and gluten feeds derived from grains not to exceed 33% of the total ration for ruminating beef and feedlot cattle older than 4 months	-	30 000 in feed ingredients and 10 000 in the total ration	-	

Grain and grain by-products not to exceed 50% of the total ration for ruminating dairy cattle older than 4 months	-	10 000 in feed ingredients and 5000 in the total ration	-
Distillers grains, brewers grains, gluten meals, and gluten feeds derived from grains not to exceed 17% of total ration for Ruminating dairy cattle older than 4 months	-	30 000 in feed ingredients and 5000 in the total ration	-
Grain and grain by-products not to exceed 50% of the total ration for chickens	-	10 000 in feed ingredients and 5000 in the total ration	-
Grain and grain by-products not to exceed 20% of the total ration for swine	-	5000 in feed ingredients and 1000 in the total ration	-
Grain and grain by-products not to exceed 40% of the total ration for all other animals	-	5000 in feed ingredients and 2000 in the total ration	-

4 Levels and distribution of mycotoxin contamination in agricultural products

Mycotoxins are among the most common natural contaminants (Smith et al., 2016). They are present in numerous commodities of plant origin. Streit et al. (2013) analyzed around 17 000 samples of both finished feed and raw feed materials such as maize, wheat, barley and soybean: 72% of samples contained detectable amounts of at least one mycotoxin. Gruber-Dorninger et al. (2019) found that 88% of around 75 000 samples collected from 100 different countries were contaminated with at least one mycotoxin. Mycotoxin contamination in food and feed is high in both developed and developing countries (Pinotti et al., 2016) with contamination levels estimated to be around 25% of the total production (Cheli et al., 2013; FAO, 2007; Ferrao et al., 2017; Hussein and Brasel, 2001; Peraica et al., 2002). Contamination occurs both during the pre-harvest period but also during transport, processing or storage with estimated annual losses of around one billion metric tons of food (FAO, 2007).

Crops are frequently contaminated by more than one type of mycotoxin. Chen et al. (2016) found co-contamination of AFB1 and ZEN in maize and its derived products. Chilaka et al. (2016) reported that 60% of maize samples were infected with at least two mycotoxins; FBs co-existed with DON in 11% of the samples. Egbuta et al. (2016) found the simultaneous occurrence of AFs and FB1 in soybean and its derived products. DON and ZEN, along with AFB1 and FBs, were commonly reported as co-existing in samples, followed by DON and FBs as well as DON, ZEN, and HT-2/T-2 toxins (Smith et al., 2016). Contamination by multiple mycotoxins is a great concern since adverse effects on animal health and performance can be additive and/or synergistic, which means that the overall toxicity is not only due to the sum but the combination of individual mycotoxins (Zachariasova et al., 2014).

The analysis of mycotoxin contamination in animal feed usually focuses on particular crops. However, silage (a mixture of grass and corn silage), fresh forage, pasture and compound feed can also be sources of mycotoxin contamination (Dell'Orto et al., 2015; Pinotti et al., 2016). Silage and compound feed are the main components of the diet, representing an average of 67% and 23% of dry matter intake, respectively. Driehuis et al. (2008) found a comparable incidence of DON in compound feed and silage. However, the same study found the average intake of DON via silage was approximately three times higher than the intake via compound feed, although the incidence of ZEN was higher in compound feed than in silage (Driehuis et al., 2008). Since compound feed is made up of a mixture of several ingredients, it is particularly vulnerable to contamination by multiple mycotoxins (Pereira et al., 2019).

The main factor affecting mycotoxin contamination is the crop species (Misihairabgwi et al., 2019). Fungal species differ in their ability to infest a particular host. Contamination also varies between varieties of the same crop (Gruber-Dorninger et al., 2019). Different types of mycotoxin may disproportionately affect different crops. Figure 8 provides examples of agricultural crop species susceptible to contamination by different groups of mycotoxins. For example, high levels of aflatoxins have been found in groundnuts and maize in Africa, Southeast Asia and southern China (WHO, 1993). Aflatoxin contamination of these crops in the USA also occurs at high levels. Although it is well known that a hot and humid climate promotes the diffusion of aflatoxin-producing molds, contamination is commonly due to the combination of meteorological conditions, environmental factors and poor agricultural practices such as incorrect harvesting and storage of crops. Data from the USA demonstrates annual and geographical fluctuations in aflatoxin levels, and increased levels in maize during droughts.

ZEN and its metabolites have been identified in many cereals including rice, sorghum, barley, oats and, particularly maize and wheat (Rai et al., 2019; Zinedine et al., 2007). OTA has been found in barley, wheat, rye, maize and rice (Shotwell et al., 1969). OTA has also been found in legumes, grapes, raisins and wines (Abarca et al., 2001; Zimmerli and Dick, 1996), nuts (Pitt, 2000), cacao and spices (Aziz et al., 1998). OTA is also found in animal products, including cow's milk and pork, particularly pork kidneys, liver and sausages (Jorgensen, 1998; Kuiper-Goodman and Scott, 1989). This shows how mycotoxins can enter the food chain and present a safety risk (Bayman and Baker, 2006).

In the case of fumonisins, the most infected crops are cereals such as rice, wheat, barley, maize, rye, oat and millet. FB1 contaminates numerous

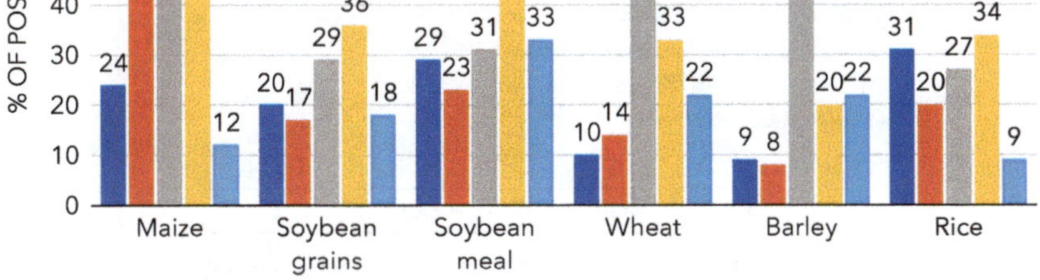

Figure 8 Incidence rate (% of positive samples) of the five main mycotoxins in different agricultural crops (adapted from Gruber-Dorninger et al., 2019).

food products, such as asparagus and garlic (Seefelder et al., 2002), dried figs (Heperkan et al., 2012) and milk (Gazzotti et al., 2009).

A second major factor in mycotoxin contamination is the climatic condition, which leads to different regional patterns of contamination by different mycotoxins. For example, high temperature, low rainfall and severe drought stress lead to the accumulation of aflatoxins. On the other hand, the *Fusarium* species producing DON and ZEN are generally associated with a cold and excessively wet growing season (Pinotti et al., 2016). Toxins occur more frequently under tropical conditions, especially in developing countries (Ferrao et al., 2017) compared to countries with colder climates, such as those in Europe (Fig. 9). However, no region of the world is completely free of mycotoxin contamination (Hesseltine, 1974).

Climate change may have profound effects on the regional pattern of mycotoxin contamination, including significant changes in atmospheric CO_2 concentrations and, in some regions, a marked decrease in summer precipitation and an increase in temperature, which would result in drought stress for some crops (Magan et al., 2011; Medina et al., 2014).

Aflatoxins continue to be a significant concern in most parts of the world. They are most often detected in Southern Europe, Africa, South and Southeast Asia (average values of positive samples higher than 30%). An incidence rate of 50% has been reported in Africa in raw cereal grains with an aflatoxin content of 1642 μg/kg. A higher incidence rate for Chinese rice has been reported

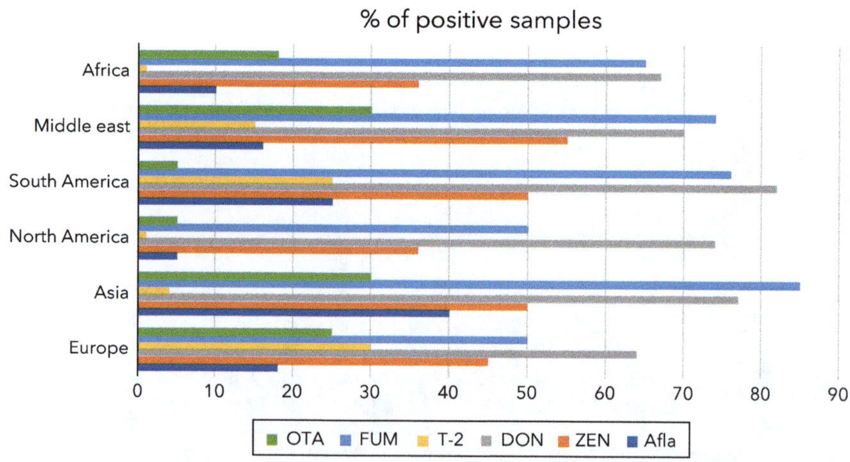

Figure 9 Mycotoxin occurrence data for each region as a percentage of all samples tested. The overall risk level for a particular region is determined by the percentage of mycotoxins that exceed the risk threshold levels for livestock (adapted from All About Feed; adapted from Gruber-Dorninger et al. (2019)).

(overall incidence at 63%) but with lower AF content. The level of contamination reported in China was 850 μg/kg, half as much as African cereals.

The highest incidence of ZEN contamination has been reported in North and South America, Central Europe, Africa and North and Southeast Asia with more than 30% of samples showing the presence of ZEN (Pinotti et al., 2016). While ZEN was not as prevalent as DON, the reported incidence in Africa and Europe was over 50%, followed by America with 48% of unprocessed cereals containing ZEN. In Asia where the incidence was the lowest at 15%, moderately high levels of ZEN (3049 μg/kg) were detected.

Higher DON concentrations (more than 60% of positive samples) have been detected in samples from temperate regions. DON is common in all regions with an average incidence of 59%, ranging from 50% in Asia to 76% in Africa, with a higher concentration in Europe and Asia. DON has been commonly found in wheat and wheat-based products with higher levels detected in wheat from China (41 157 μg/kg).

The highest OTA prevalence has been reported in South Asia and Africa with more than 50% of positive samples (Pinotti et al., 2016). The highest level of OTA (1164 μg/kg) was detected in rice samples from Africa. Rice is an aquatic plant and is usually harvested at very high moisture levels ranging between 35% and 50%, making it vulnerable to OTA contamination (Duarte et al., 2010; Zinedine et al., 2007).

The highest incidence of fumonisin contamination (more than 50% of positive samples) has been found in South America, Southern Europe, Africa, North, South and Southeast Asia (Gruber-Dorninger et al., 2019; Ottoboni et al., 2016; Pinotti et al., 2016). The highest incidence rate of 95% has been reported in South America with a level of contamination ranging between 10 μg/kg and 34 700 μg/kg in raw cereal grains. An incidence rate of 62% has been reported in both Asia and Africa with the highest levels being 17 121 μg/kg and 24 225 μg/kg, respectively. Both the lowest incidence (39%) and concentration (5400 μ/kg) in the samples tested have been reported in Europe.

Geographical area, seasonal and local weather conditions during critical plant growth stages (before, during flowering or in grain at maturity) are of great importance in explaining variations from different surveys (Gruber-Dorninger et al., 2019; Pinotti et al., 2016; Streit et al., 2013). For example, high rainfall during the silking period facilitates mycotoxin infestation of plants, which contributes to explaining the differences in contamination between years in the same region (Gruber-Dorninger et al., 2019).

The significant differences in type and prevalence of mycotoxin contamination in different regions of the world are shown in Fig. 10 (Pinotti et al., 2016). The geographical distribution of mycotoxin contamination can be illustrated in a map of the global risk of mycotoxins (Fig. 11). This map shows that Europe ranks as a severe risk region, with more than half of the samples

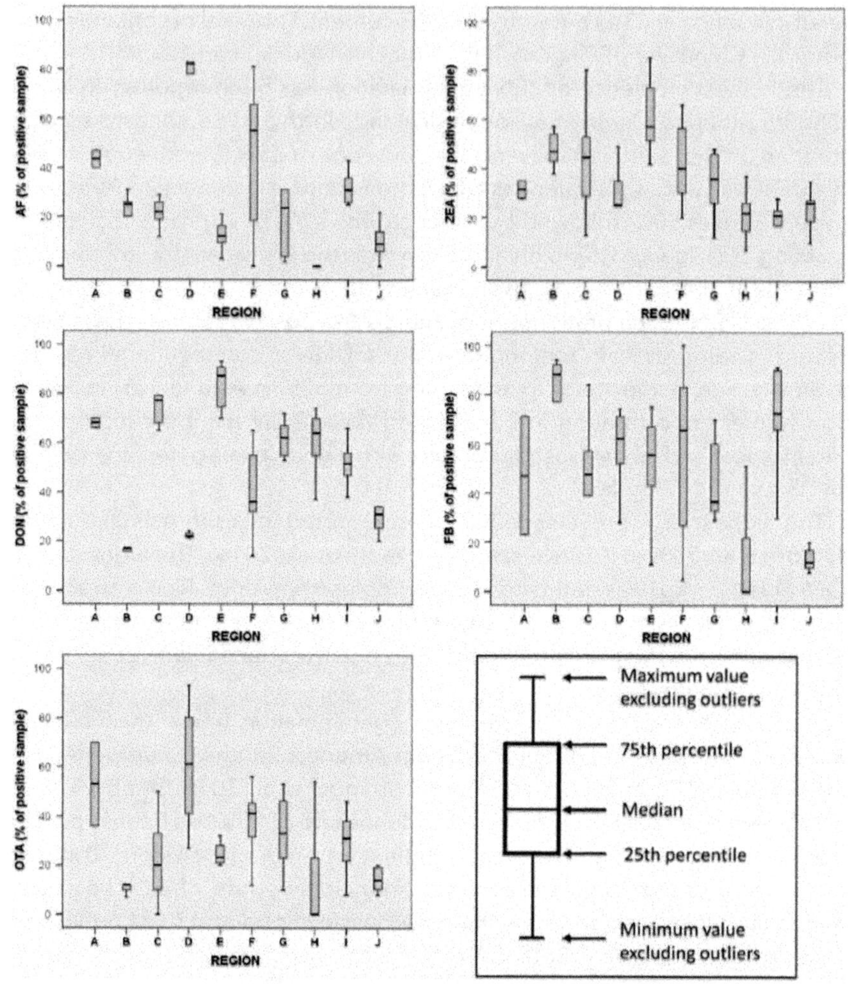

Figure 10 Global mycotoxin occurrence (percentage of mycotoxin-contaminated samples) in raw feed materials and finished feed. A, Africa; B, South America; C, North America; D, South Asia; E, North Asia; F, Southeast Asia; G, Central Europe; H, North Europe; I, South Europe; J, Oceania. The figure is taken from Pinotti et al. (2016).

tested being above risk threshold levels. The highest incidence was DON with 65% of samples showing an average of 555 ppb. The second most prevalent mycotoxin was FUM, present in 51% of samples. ZEN occurred in 44% of samples, with the highest single level in an Austrian corn sample (6082 ppb) (All About Feed, 2018). The T-2 toxin showed lower values, with an incidence rate of 33%.

The Americas pose extreme/severe mycotoxin-related risks with more than 80% and 71% of tested samples being above risk threshold levels for North

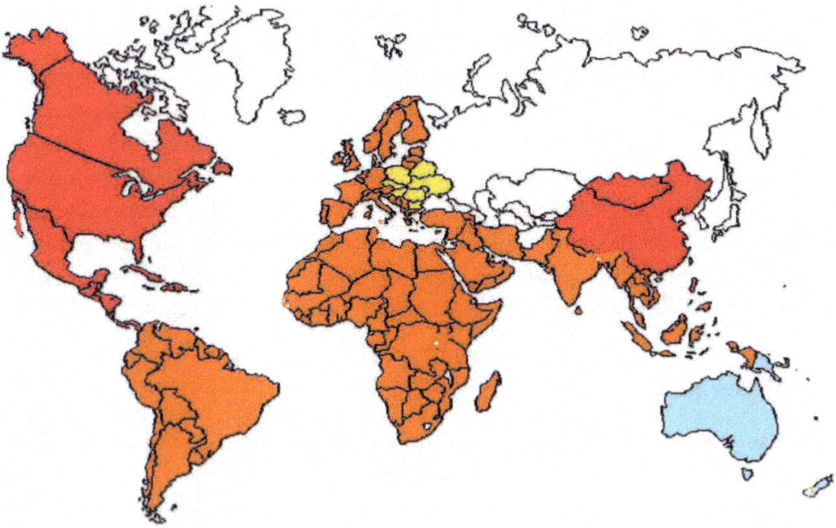

Figure 11 Risk map. Red represents a very high risk; orange severe risk; yellow high risk; light blue low risk (adapted from All About Feed, 2018 and Gruber-Dorninger et al, 2019).

and South America, respectively. The most frequently occurring mycotoxin was DON, detected in 75% of the samples with an average concentration of 955 ppb. FUM, ZEN, aflatoxins and OTA were detected in 52%, 37%, 6% and 5% of samples, respectively, with average levels of 2 ppb, 41 ppb, 213 ppb, 36 ppb and 4 ppb, respectively (All About Feed, 2018).

5 Mycotoxicosis symptoms in animals

Mycotoxicosis is a disease caused by the ingestion of grains or forage containing toxic metabolites produced by certain fungi. Mycotoxins are considered a serious risk to animal health (Khaneghah et al., 2019). Most exposure is probably chronic or subacute as a result of consuming small amounts of toxins over a long period of time. In these instances, the principal signs of toxicosis are the impairment of metabolic, nutritional or endocrine functions. Symptoms include decreased appetite, poor growth rate, reduced average daily feed intake and feed efficiency, decreased reproductive performance, immunosuppression, increasing susceptibility to secondary diseases, teratogenicity, and cancer. However, diagnosis of chronic mycotoxicosis is often difficult because the clinical signs are seldom overt and lesions are not specific (Bryden, 2012; EFSA, 2017; FAO, 2007; https://vetmed.iastate.edu/vdpam/FSVD/swine/index-diseases/mycotoxicoses). Due to regulatory limits in many countries, mycotoxin contamination levels in animal feedstuffs are usually not high enough to cause acute illness. Acute outbreaks may have more obvious signs and vary depending on the different mycotoxins (https://vetmed.iastate.edu/vdpam/FSVD/swine/index-diseases/mycotoxicoses).

The rate of toxicity depends on a range of factors, such as the toxin present, dosage, duration of exposure, animal species, age, hormonal status, nutrition, gut microflora status and current disease (Bryden, 2007). The particular species affect the toxicokinetics, bioavailability and mechanisms of action of mycotoxins (Bertero et al., 2018). Table 3 summarizes the main effects of different mycotoxins on different types of livestock. Of the over 400 mycotoxins identified to date, six have been reported to cause diseases in animals. These are commonly called the 'big six' and are AFB1, ZEN, DON, OTA, T-2 toxin and FUM.

5.1 Aflatoxins

AFB1 is the most potent natural carcinogen. It contaminates crops such as cereals, figs, oilseeds and nuts; contamination can occur during storage in conditions such as high humidity (Bennett and Klich, 2003). It is rapidly absorbed and metabolized primarily in the liver in the microsomal system. Mutagenic action is linked to the production of an epoxide, an activated metabolite that forms a covalent bond with DNA nucleic acids. This bonding inhibits the transcription process responsible for DNA repair as well as producing lesions, mutations and subsequent tumor formation such as hepatocellular carcinoma (Hussein and Brasel, 2001).

In horses, aflatoxicosis can cause poor appetite, anorexia, fever, depression, lameness, damage to skeletal muscles (tremor, ataxia), damage to the liver and heart and even death (Caloni et al., 2011; Vesonder et al., 1991). AFB1 contamination in pigs reduces the average daily weight gain and consequently causes a reduction in the overall growth rate (Harvey et al., 1988; Huff et al., 1988; Southern and Clawson, 1979). The major effect of aflatoxins on poultry is to reduce feed intake and body weight with other effects on layers (Garlich et al., 1973), such as small egg size (Bryden et al., 2012) and less likelihood of hatchability (Howarth and Wyatt, 1976). In pets, symptoms of acute toxicoses are hepatitis, vomiting, depression, polydipsia and polyuria. LD50 levels ranging from 0.5 mg/kg to 1 mg/kg in dogs and 0.3 mg/kg to 0.6 mg/kg in cats can be fatal within a few days (Hussein and Brasel, 2001).

Aflatoxins in ruminants are only partially degraded by the ruminal flora and, because of the ruminant's metabolism, a secondary metabolite (aflatoxicol) is produced. Exposure to aflatoxins results in negative effects such as poor immune system function. Rumen metabolism can also be affected, which can then reduce cellulose digestion, volatile fatty acid production and rumen motility, as well as reducing feed intake and milk quality (Choudhary et al., 1998; Cook et al., 1986; Helferich et al., 1986; Paul et al., 1977).

Aflatoxin B1 (AFM1) bypasses rumen degradation and is partially converted by hepatic metabolism into aflatoxin M1. It is subsequently excreted via bile or enters systemic circulation (Fink-Gremmels, 2008a,b). Aflatoxin M1 can therefore be excreted through urine or in milk. The transfer rate of AFM1 varies between 1% and 6% (EFSA, 2004). Excretion in milk (carryover) varies according to nutritional and physiological factors, including feeding regime, rate of ingestion and digestion, the health of the animal, hepatic biotransformation capacity, lactation status, and the number of milkings after exposure (Fink-Gremmels, 2008b). This means that the rate of absorption of aflatoxins, and the excretion of aflatoxin M1 in milk, varies between individual animals, from day to day, and from one milking to the next. In high-yielding cows, the consumption of significantly higher amounts of concentrated feeds could result in carryover percentages as high as 6.2% (Fink-Gremmels, 2008b).

5.2 Zearalenone

Zearalenone is a lactone non-steroidal estrogenic mycotoxin produced by several *Fusarium* species, above all *F. graminearum* (Iheshiulor et al., 2011; Lee et al., 2018). ZEN is rapidly well absorbed after oral administration and can undergo prehepatic, hepatic and extrahepatic metabolism (Bertero et al., 2018). The predominant metabolic route and the number of its metabolites are one of the reasons why different sensitivities to this toxin are observed among animal species (Bertero et al., 2018). It has strong estrogenic activities because it mimics the action of the female hormone, 17β-estradiol (Bryden et al., 2012). It thus acts as an endocrine disruptor, causing alteration in the reproductive tract, and can interfere with ovulation, conception, implantation and fetal development (Iheshiulor et al., 2011).

Pigs, in particular prepubertal females, are considered the most sensitive species because ZEN can interfere with the reproductive system and alter the mammary gland (Alm et al., 2002; Cortinovis et al., 2014; Dai et al., 2016; Ranzenigo et al., 2008). The higher sensitivity to ZEN is probably due to pigs being able to convert ZEN into a more estrogenically active molecule, α-zearalenol (Minervini and Dell'Aquila, 2008). In pregnant gilts and sows, ZEN can increase the chances of a decreased fertilization rate, abortion, stillbirths and ovarian cysts (Glavitis and Vanyi, 1995). In contrast, ruminants seem less susceptible to ZEN toxicity, although severe symptoms such as loss of reproductive ability have been described in calves and young heifers (Bertero et al., 2018; Pizzo et al., 2016). This is because, in young animals, the rumen is not completely functional and is not able to convert this mycotoxin into a less active form (Fink-Gremmels, 2008a).

5.3 Deoxynivalenol

Deoxynivalenol, also known as vomitoxin, is a member of the trichothecene mycotoxin family and commonly contaminates the kernels of infected

Table 3 Effects of the six main mycotoxins

Mycotoxins	Species susceptibility	General effects	Species-specific effect: pigs	Poultry	Cows	Horses
Aflatoxins B1	All domestic animals and poultry	Hepatoxicity, carcinogenic, immunosuppressive, decrease in productive parameters, gastrointestinal dysfunction, anemia, hemorrhage	Hepatic disorders, reduced weight gain, inferior carcass quality, abortions	Hepatotoxic, reduction in egg production, embryotoxicity	Reduced production, alteration in rumen activity	Hepatic damage
Zearalenone	Mainly pigs and dairy animals	Estrogenic and reproductive disorders	Reddening and swelling of the vulva and mammary glands, vaginal prolapse, decline in sow prolificacy, reduction in quantity and quality of sperm, pseudopregnancy feminization of young males	Poor fertility and hatchability, poor growth of progeny, the decline in egg production	Embryo death, numerous returns to heat, ovarian cysts	
Deoxynivalenol	Mainly pigs and dairy animals	Dermatotoxic, feed refusal	Nervosity, loss of appetite, deterioration in feed efficiency, diarrhea, vomiting, refusal to eat	Intestinal modification	Poor assimilation of feed, the decline in production and quality of milk, immune system disorders	
Ochratoxin	Mainly pigs and poultry	Nephrotoxic, gout	Nephropathy, anorexia, faintness, uncoordinated movements	Teratogen in poultry embryo, reduced growth, decreased feed conversion ratio and egg production, feed refusal	Significant immune system and carcinogenic disorders	

T-2 toxin	Mainly pigs and poultry	Lower feed intake, loss of appetite, decreased weight gain, low production	Loss of appetite, intestinal disorders (enteritis), reduced reproductive performance, immunodepression	Lower feed intake, reduced weight gain, lower egg production, necrotic lesion in mouth, crop, gizzard, intestinal mucosa	Loss of appetite, reduced reproductive performance	
Fumonisins	Mainly pigs and horses	Neurological disorders, liver damage	Hepatopathie, pulmonary edema	Decreased body weight, liver pathology	Peak mortality, reduced feed intake, poor growth.	ELEM

cereals, especially wheat and wheat products, rendering the grain unsuitable for human or animal consumption (Foroud and Eudes, 2009). In general, trichothecenes are easily absorbed via the tegumentary and gastrointestinal systems. However, ingestion can cause intestinal irritation in mammals and can lead to feed refusal in livestock. Trichothecenes inhibit eukaryotic protein synthesis, specifically by preventing peptide bond formation, although the polypeptide chain termination may also be inhibited (Bertero et al., 2018). The main cellular targets of trichothecenes are leucocytes. These mycotoxins can thus cause immunosuppression in species such as swine, although other species, such as horses, are less affected (Bertero et al., 2018; Reddy et al., 2018). These mycotoxins are also potential endocrine disruptors with effects on follicular maturation and steroidogenesis in swine, thus impacting on the reproductive performance (Gerez et al., 2017; Ranzenigo et al., 2008).

Acute toxicity is characterized by gastrointestinal signs such as abdominal discomfort, diarrhea, vomiting and anorexia. The most common clinical signs of chronic exposure in swine are weight loss, anorexia, decreased nutritional efficiency and immune dysregulation due to a neurochemical imbalance in the porcine brain (Bertero et al., 2018; Bryden, 2012; Pestka, 2007). Other clinical signs are lesions in the stomach, intestine, lungs and kidneys (Bertero et al., 2018). Pigs, and in particular young pigs, are poorly tolerant to DON and are considered the most sensitive species (Pestka, 2007).

Bovines are less sensitive to DON than pigs because of their rumen microbe metabolism (Fink-Gremmels, 2008a). Only small traces of DON (<1%) are absorbed and found in the systemic circulation, indicating a nearly complete degradation of the molecule in the rumen. Rumen microflora transforms DON into the less toxic metabolite DOM-1 (Bertero et al., 2018). However, where a ruminal system is less efficient, such as ruminants with acidosis or young animals, the toxicokinetics of DON may differ and these animals can be more susceptible to the toxic effects of this mycotoxin (Bertero et al., 2018). While deoxynivalenol does not affect milk production, the milk fat content decreases, and the immune and reproductive systems are altered (Charmley et al., 1993; Korosteleva et al., 2009; Pizzo et al., 2015).

Poultry seems to have a greater tolerance to high doses of DON in terms of performance and productivity compared to other species. They have low levels of DON absorption in plasma together with rapid metabolism and clearance. Intestinal microflora can convert DON to DOM-1 (Bertero et al., 2018). This mycotoxin alters the small intestinal morphology but has no effect on egg yield, weight and shell thickness (Awad et al., 2011; Bergsjø et al., 2007; Sypecka et al., 2004). In pets, high levels of DON in food cause reduced food consumption, food refusal and vomiting.

5.4 T-2 and HT-2 toxins

Type A trichothecenes include T-2 toxin and HT-2 toxin. These mycotoxins are produced by various *Fusarium* species which infect crops in the field or during storage (Bertero et al., 2018). T-2 toxin affects the nervous system, including limb position and reflexes. Moreover, it also affects the immune system and shows immunomodulatory activity in a time- and dose-dependent manner (Bertero et al., 2018; Iheshiulor et al., 2011). The main effects of trichothecene mycotoxins are reduced feed intake and weight gain, gastrointestinal disorders, neuroendocrine and hepatological changes, and impairment of the immune system. Moreover, T-2 toxin can cause serious hemorrhagic inflammation by damaging the blood vessel walls (Bertero et al., 2018).

Pigs are the most susceptible animals. Contamination mainly reduces reproductive performance and causes weight loss (Caloni et al., 2009). Immunological and/or hematological effects can also occur in pigs (Meissonnier et al., 2008). The intestinal tract of chickens quickly absorbs, metabolizes and largely eliminates T-2 toxin (by about 90%). The first signs of T-2 toxicosis in poultry are lower feed intake, reduced weight gain and growth retardation, lower egg production and egg weight, thinner eggshells, more fragile shells and decreased hatchability (Chi et al., 1977; Tobias et al., 1992; Wyatt et al., 1975). Ruminants are considered less sensitive to the effects of the T-2 toxin thanks to detoxification by the rumen microflora. In the ruminal fluid, T-2 toxin is rapidly degraded into various metabolites, all of which lack significant toxicity (Fink-Gremmels, 2008a). Young animals are thus more susceptible to this toxin due to the incomplete functionality of the rumen.

5.5 Ochratoxins

OTA is the main metabolite of toxicological significance for livestock (Iheshiulor et al., 2011; Reddy and Bhoola, 2010), resulting from biotransformation due to cytochrome P450 which occurs both in humans and animals. Its absorption takes place in the gastrointestinal tract and, once in the blood, it binds to albumin (Kőszegi and Poór, 2016) which results in a long residence time in the body.

The target organs for ochratoxins are the kidneys, which cause nephropathy in all species, especially in swine (Krogh, 1991). OTAs have been associated with endemic nephropathy, liver damage and necrosis of intestinal and lymphoid tissue in swine (Pereira et al., 2019). Ochratoxins do not cause significant toxicity to cattle probably because of ruminal microbial degradation and detoxification (Patterson et al., 1981). OTA is rapidly deactivated by the forestomach flora, and only very small amounts of intact OTA are absorbed (Fink-Gremmels, 2008a).

5.6 Fumonisin B1

The mechanism of action of Fumonisin B1 is based on the inhibition of the enzyme ceramide synthase, which causes ceramide biosynthesis (acylation of sphinganine and sphingosine), with the accumulation of intermediates of sphingolipid biosynthesis (sphinganine). Consequently, the sphingolipid metabolism in tissue is altered and the accumulation in sphinganine tissue initiates a cascade of events that can cause toxicity, especially in the liver and kidneys or lead to loss of cell membrane integrity and structure, causing carcinogenicity (Bertero et al., 2018; Kamle et al., 2019).

Horses are the most sensitive species to FB1 toxicity, with their central nervous system, liver and heart being the target organs (Bertero et al., 2018). In this species, FB1 appears in a neurotoxic (leukoencephalomalacia-ELEM) form which occurs rapidly with high mortality (Ross et al., 1991; Vendruscolo et al., 2016; Thiel et al., 1991).

Along with horses, pigs are the most sensitive species to fumonisins. A typical presentation of acute intoxication is a huge pulmonary edema called a 'porcine pulmonary edema' (PPE) (Colvin et al., 1993). This is due to an increase in pulmonary capillary hydrostatic pressure because of acute left-side heart failure and an increase in vascular permeability which is caused by damage to the alveolar endothelium and alveolar epithelium (Bertero et al., 2018). Typical clinical signs include reduced feed consumption, dyspnea, weakness, cyanosis, and death (Pereira et al., 2019). Other signs of fumonisin toxicosis are liver problems (Colvin et al., 1993), cardiovascular disease (Bertero et al., 2018), immune system damage (Pierron et al., 2016), and poor reproductive performance (Cortinovis et al., 2014). Hepatic damage (characterized by cholestasis, hepatic necrosis, cell proliferation) is dose- and time-dependent. It is marked by an increase in the activity of the hepatic enzymes in serum, icterus, hepatocellular necrosis, anorexia and hepatic encephalopathy (Bertero et al., 2018; Colvin et al., 1993).

Ruminants are less sensitive to FB1 than monogastric species because of the low bioavailability after ingestion and the ability of FB1 to pass the rumen without microbial degradation (Bertero et al., 2018; Caloni et al., 2000). Poultry are also less susceptible and are considered relatively resistant to the toxic effects of FB1 (Murugesan et al., 2015). However, a decrease in body weight and liver pathology has been observed (Brown et al, 1992; Ledoux et al., 1992).

6 Mycotoxin detection techniques

The analysis of mycotoxin compounds is very complex because of the number of different types, variable chemical structures, as well as differing rates of diffusion in various matrices, depending on the class of mycotoxin. Furthermore, little is known about many mycotoxins and therefore they are difficult to detect. The most common mycotoxins such as aflatoxin B1 and DON are considered as indicators: if they are found in the sample, other mycotoxins are also likely to be present. The choice of technique depends on the speed of analysis, whether there is a need to determine one or more mycotoxins and the applicability to different matrices.

6.1 Screening methods

Mycotoxin screening methods are available as kits, the most popular being immunological assays, such as enzyme-linked immunosorbent assay (ELISA). The principle is based on the identification of a particular antibody by binding with the specific antigen. The mycotoxin acts as an antigen on the respective antibody and the detection of the bond is achieved by exploiting typical markers that are radioactive, fluorescent or chromogenic. When using this method, it is the intensity of the color that determines the diffusion of the bond. The characteristics of this method are:

- good specificity (for a given mycotoxin or group of mycotoxins),
- good sensitivity (in measuring ppt),
- good reliability of results, and
- speed and low cost.

Kits vary in terms of factors such as the matrix to be analyzed, the sensitivity required and the speed of analysis. The main problem with immunological assays is the inability to discriminate between mycotoxin subtypes, for example, aflatoxin B1 and B2.

6.2 Chromatographic methods

Positive samples from immunological tests can be confirmed using chromatography. This technique has the advantages of high sensitivity (especially when combined with mass spectrometry), specificity (including identifying subtypes of toxins), accuracy and reproducibility of results. It is used as a control technique because it requires expensive equipment, qualified personnel and can only analyze one sample at a time. While it is unsuitable for screening, chromatography is ideal for the validation of other tests.

High-performance liquid chromatography (HPLC) is the most commonly used technique, although gas-chromatography (GC) is also used. LC is affected by the sample type, preparation and matrix effect, which means that each sample needs to be prepared and analyzed separately. The pre-treatment technology, Quick, Easy, Cheap, Effective, Rugged and Safe (QuEChERS) was originally developed in 2003 for pesticide determination. This process is an extraction step using acetonitrile and salts, followed by a clean-up step based on a dispersive solid-phase extraction (dSPE). Due to the operating simplicity involved, QuEChERS-based approaches have been used increasingly in the field of multi-mycotoxin analysis in feed/food samples. The QuEChERS procedure is now widely used in routine mycotoxin analysis and is also useful for emerging mycotoxins.

A wide range of chromatography methods have been developed combining LC with mass spectrometry (MS) (LC-MS). The current 'gold' standard in laboratories concerned with food/feed safety control is the unit resolution tandem mass spectrometric detector (MS/MS). A mass spectrometer works by ionizing molecules, which it then sorts and identifies based on their mass-to-charge ratio (*m/z*). MS detectors offer higher sensitivity and selectivity, as well as chemical structural information on the analyte based on *m/z* ratio, making them an ideal confirmation technique. For this reason, mass spectrometry is increasingly paired with HPLC (HPLC-MS).

Other methods used to identify toxins require further verification and validation by organizations, such as the ISO, and include infrared (Levasseur-Garcia, 2018) and Raman spectroscopy (Neng et al., 2020). Methods involving infrared analyzers enable even faster analysis because they do not require sample preparation. Both techniques are used, for example, for the analysis of the DON contained in wheat and corn.

A new frontier in mycotoxin detection is the combination of immunological essays with other techniques such as electronic noses, which consist of an array of non-specific chemical detectors that detect different volatile organic compounds (VOCs). The assumption is that the growth and biochemical pattern of mycotoxigenic fungi species cause chemical changes in the composition of VOCs (Olsson et al., 2002). E-noses are used to identify specific VOCs associated with the growth of fungi on cereal grains or other feed/food (Ottoboni et al., 2018).

7 Mycotoxin prevention and control

The agri-food sector, including the animal feed industry, is highly sensitive to the issue of mycotoxins, both from an economic and safety point of view (Agriopoulou et al., 2020). Various techniques have been developed to prevent or treat contamination, both in the field and/or in the subsequent pre-harvest, post-harvest, transport, storage and processing phases of feed production.

Prevention is the main way for limiting contamination, including the use of fungus-resistant varieties and good agricultural practices to prevent the conditions favorable to mold development. However, due to the huge difficulty in avoiding crop contamination directly in the field (e.g. because of changing weather conditions), the use of post-harvest technological treatments, such as disinfection of crops, and storage at low temperatures and in a controlled atmosphere (Agriopoulou et al., 2020), is the most common and effective ways to prevent crop contamination (Cheli et al., 2017; Čolović et al., 2019).

To prevent mycotoxins from becoming a health hazard, manufacturers are required to conduct hazard analyses that identify the potential hazards within the manufacturing process and provide evidence of consistent monitoring and preventive controls for mycotoxins in animal feed (https://www.fda.gov/animal-veterinary/biological-chemical-and-physical-contaminants-animal-food/chemical-hazards). Effective monitoring and control require the implementation of hazard analysis and critical control point (HACCP) approaches, good manufacturing practices, and quality control systems. Processors need, for example, to verify adequate storage conditions in grain silos with optimal temperature and humidity to prevent toxicogenic mold from developing (Peraica et al., 2002). Implementing a comprehensive and effective mycotoxin testing program is also critical.

7.1 Mechanical treatments

Mechanical processes do not destroy mycotoxins but vary their distribution and concentration in the final product. Sorting, cleaning and grinding cereal grains seem to be effective in reducing mycotoxins (Pascale et al., 2020). Cleaning the grain removes mycotoxins since they can accumulate in the dust between grains which is removed during the cleaning process. Sorting can also eliminate mold-affected grains. The removal of contaminated grains from bulk leads to a reduction in aflatoxins in corn by 40–80%, while wheat washing and removal of dust or other impurities can reduce DON by 29–90% and T-2 and HT-2 toxins by 25–80%. However, the effectiveness of these operations depends on, for example, the initial level of contamination of the raw material.

Similarly, milling redistributes mycotoxins rather than destroying them. Not all parts of the grain are equally susceptible to mold attack: the outer part of the seed, as well as the germ and pericarpal, are more susceptible than the endosperm. As a result, mycotoxin levels are generally lower in fractions intended for human consumption (flour or semolina), and higher in fractions mainly intended for animal feeds, which are composed of those grain by-product parts most susceptible to contamination. In the case of wheat, the understanding of mycotoxin repartitioning in milling fractions is largely limited to DON.

7.2 Thermal treatments

Thermal processes, such as pelleting and extrusion, are widely used in the production of feed. These processes also affect the number of mycotoxins in the matrix. However, their impact can be limited because most mycotoxins are stable at high temperatures (Čolović et al., 2019). The most important factors influencing reduction are the type of mycotoxin, its initial concentration, and the conditions under which the thermal treatment takes place. Temperature, exposure time and the degree of penetration of heat into the matrix also influence mycotoxin reduction. Extrusion shows the highest potential due to the combination of high temperatures, pressure and the sheer force used. Kabak et al. (2009) reported that extrusion at temperatures above 150°C significantly reduces ZEN and fumonisins. The extrusion process is very effective with fumonisins, as confirmed by Costelo et al. (1998). At temperatures above 160°C, the fumonisin content is reduced by 46–92% while, at temperatures between 70°C and 105°C, fumonisin content is reduced by less than 30%. At the same higher temperatures, however, there is only a slight reduction in other toxins, such as aflatoxins and deoxynivalenol.

Overheated steam is another thermal technology that is effective in reducing toxins with less loss of oxidation and greater energy efficiency (Anto et al., 2014). Pronyk et al. (2006) studied the effects of DON reduction in grains using an overheated steam process, focusing on the duration and temperature (experimental range: 110–185°C). At 185°C, DON was reduced by up to 50%, but lowering the temperature to 135°C had no significant effects. It is important to bear in mind the speed at which steam reaches the matrix: too high or too low a speed can achieve the opposite effects, regardless of temperature.

A common dilemma with all thermal treatments is to find the right balance between time and operating temperature so as not to damage the product. If it is important to reduce the content of mycotoxins in the matrix, it is equally important to protect nutrients from thermal degradation. Many components are thermosensitive, such as starch which can be gelatinized and vitamins which are destroyed at high temperatures, reducing the nutritional value of the product.

7.3 Ensiling

Ensiling is a classic forage conservation method for ruminants which can be used as a post-harvest method to prevent the formation of mycotoxins (Dell'Orto et al., 2015). The anaerobic activity of the lactic acid bacteria in the forage leads to a change in pH and in the activity of enzymes including proteases, amylases and hemicellulases. The resulting fermentation eliminates mycotoxin-producing spoilage molds as well as improving the palatability and nutritional profile of the feed (Adebiyi et al., 2019).

8 Conclusion

Mycotoxins are toxic metabolites produced by certain fungi (molds) which can infect agricultural products both in the field and/or during storage.

The occurrence of these toxins in grains and other animal feed materials is influenced by environmental factors such as temperature, humidity level, and/or rainfall during growing, harvesting, and post-harvesting. If infestation occurs, mycotoxins are produced in feed products. When consumed by animals, mycotoxins can cause chronic toxicity or even death. Mycotoxins can lead to various toxicological problems such as neurological issues, liver failure, kidney failure, and cardiac failure (https://www.fda.gov/animal-veterinary/biological-chemical-and-physical-contaminants-animal-food/chemical-hazards). All these issues affect the mycotoxins studied to date, in particular, the 'big six'. However, other 'emerging mycotoxins', 'masked mycotoxins' and the effects of co-occurrence of multiple mycotoxins need to be further investigated.

In the future, it will be necessary to combine feed and animal health monitoring to measure the actual levels of mycotoxin contamination and their impact on animals. Nutritionists and veterinarians also need to optimize animal health since, if animal tolerance to disease is reduced, even a low level of mycotoxins in feed can pose a threat to their productivity and health. More effective risk assessments may make it possible to redirect contaminated feed to less vulnerable animal species. It may also be possible to use feed additives that suppress or reduce the absorption and/or promote the excretion of mycotoxins or modify their action (Pereira et al, 2019).

9 Where to look for further information

Weidenbörner, M. (2007). *Mycotoxins in Feedstuffs*. Springer, New York, NY gives a detailed and more thorough introduction on mycotoxins and their occurrence. For further information regarding detection and/or analyses see:

- Magan, N. and Olsen, M. (2004). *Mycotoxins in Food: Detection and Control*. CRC Press, Cambridge, UK.
- Moretti, A. and Susca, A. (2017). *Mycotoxigenic Fungi*. Humana Press, New York, NY.
- Barug, D., Bhatnagar, D., van Egmond, H. P., van der Kamp, J. W., van Osenbruggen, W. A. and Visconti, A. (2006). *The Mycotoxin Factbook: Food & Feed Topics*. Wageningen Academic Publishers, Wageningen, Netherlands.
- De Saeger, S. (2011). *Determining Mycotoxins and Mycotoxigenic Fungi in Food and Feed*. Woodhead Publishing, Cambridge, UK.

Since mycotoxins can appear in the food chain as a result of fungal infection of crops, problems related to human health and further information can be found at the following websites:

- World Health Organization (https://www.who.int/news-room/fact-sheets/detail/mycotoxins).
- European Food Safety Authority (https://www.efsa.europa.eu/en/topics/topic/mycotoxins).
- U.S. Food and Drug Administration (https://www.fda.gov/animal-veterinary/biological-chemical-and-physical-contaminants-animal-food/chemical-hazards#Mycotoxins).

Concerning animal health issues and feed production, a more detailed description of the problems are available at:

- www.fao.org/3/Y1390E/y1390e02.htm.
- IFIF-FAO Feed Manual 2020 on good practices for the feed sector. Available online: https://ifif.org/our-work/education-best-practices.
- https://www.allaboutfeed.net/Mycotoxins/Keep-up-to-date-on-mycotoxins/.

10 References

Abarca, M. L., Accensi, F., Bragulat, M. R. and Cabañes, F. J. (2001). Current importance of ochratoxin A-producing Aspergillus spp, *J. Food Prot.* 64(6), 903–906.

Agriopoulou, S., Stamatelopoulou, E. and Varzakas, T. (2020). Advances in occurrence, importance, and mycotoxin control strategies: prevention and detoxification in foods, *Foods* 9(2), 137–184.

All about Feed (2018). Available at: https://www.allaboutfeed.net/Mycotoxins/Keep-up-to-date-on-mycotoxins/.

Alm, H., Greising, T., Brussow, K. P., Torner, H. and Tiemann, U. (2002). The influence of the mycotoxins deoxynivalenol and zearalenol on in vitro maturation of pig oocytes and in vitro culture of pig zygotes, *Toxicol. In Vitro* 16(6), 643–648.

Alshannaq, A. and Yu, J. H. (2017). Occurrence, toxicity, and analysis of major mycotoxins in food, *Int. J. Environ. Res. Public Health* 14(6), 632–651.

Awad, W. A., Hess, M., Twarüzek, M., Grajewski, J., Kosicki, R., Böhm, J. and Zentek, J. (2011). The impact of the Fusarium mycotoxin deoxynivalenol on the health and performance of broiler chickens, *Int. J. Mol. Sci.* 12(11), 7996–8012.

Aziz, N. H., Youssef, Y. A., El-Fouly, M. Z. and Moussa, L. A. (1998). Contamination of some common medicinal plant samples and spices by fungi and their mycotoxins, *Bot. Bull. Acad. Sin.* 39, 279–285.

Bayman, P. and Baker, J. L. (2006). Ochratoxins: a global perspective, *Mycopathologia* 162(3), 215–223.

Bennett, J. W. and Klich, M. (2003). Mycotoxins, *Clin. Microbiol. Rev.* 16(3), 497–516.

Bergsjø, B., Herstad, O. and Nafstad, I. (2007). Effects of feeding deoxynivalenol-contaminated oats on reproduction performance in white leghorn hens, *Br. Poult. Sci.* 34, 147–159.

Bertero, A., Moretti, A., Spicer, L. J. and Caloni, F. (2018). Fusarium molds and mycotoxins: potential species-specific effects, *Toxins* 10(6), 244–270.

Bordin, K., Sawada, M. M., da Costa Rodrigues, C. EdC., da Fonseca, C. R. and Oliveira, C. A. F. (2014). Incidence of aflatoxins in oil seeds and possible transfer to oil: a review, *Food Eng. Rev.* 6, 20–28.

Brown, T. P., Rottinghaus, G. E. and Williams, M. E. (1992). Fumonisin mycotoxicosis in broilers: performance and pathology, *Avian Dis.* 36(2), 450–454.

Bryden, W. L. (2012). Mycotoxin contamination of feed supply chain: implication for animal productivity and feed security, *Anim. Feed Sci. Technol.* 173(1–2), 134–158.

Caloni, F. and Cortinovis, C. (2010). Effects of fusariotoxins in the equine species, *Vet. J.* 186(2), 157–161.

Caloni, F., Spotti, M., Auerbach, H., Op den Camp, H., Gremmels, J. F. and Pompa, G. (2000). In vitro metabolism of fumonisin B1 by ruminal microflora, *Vet. Res. Commun.* 24(6), 379–387.

Caloni, F., Ranzenigo, G., Cremonesi, F. and Spicer, L. J. (2009). Effects of a trichothecene, T-2 toxin, on proliferation and steroid production by porcine granulosa cells, *Toxicon* 54(3), 337–344.

Charmley, E., Trenholm, H. L., Thompson, B. K., Vudathala, D., Nicholson, J. W., Prelusky, D. B. and Charmley, L. L. (1993). Influence of level of deoxynivalenol in the diet of dairy cows on feed intake, milk production, and its composition, *J. Dairy Sci.* 76(11), 3580–3587.

Cheli, F., Gallo, R., Battaglia, D. and Dell'Orto, V. (2013). EU legislation on feed related issues: an update, *Ital. J. Anim. Sci.* 12(2), 295–312.

Cheli, F., Battaglia, D., Gallo, R. and Dell'Orto, V. (2014). EU legislation on cereal safety: an update with a focus on mycotoxins, *Food Control* 37, 315–325.

Cheli, F., Pinotti, L., Novacco, M., Ottoboni, M., Tretola, M. and Dell'Orto, V. (2017). Mycotoxins in wheat and mitigation measures. In: Wheat *Improvement, Management* and *Utilization*. London: IntechOpen, 227–251.

Chen, Y., Chen, Q., Han, M., Zhou, J., Gong, L., Niu, Y., Zhang, Y., He, L. and Zhang, L. (2016). Development and optimization of a multiplex lateral flow immunoassay for the simultaneous determination of three mycotoxins in corn, rice and peanut, *Food Chem.* 213, 478–484.

Chi, M. S., Mirocha, C. J., Kurtz, H. F., Weaver, G., Bates, F. and Shimoda, W. (1977). Effects of T-2 toxin on reproductive performance and health of laying hens, *Poult. Sci.* 56(2), 628–637.

Chilaka, C. A., De Boevre, M., Atanda, O. O. and De Saeger, S. (2016). Occurrence of Fusarium mycotoxins in cereal crops and processed products (Ogi) from Nigeria, *Toxins* 8(11), 342–359.

Choudhary, P. L., Sharma, R. S., Borkhataria, V. N. and Desai, M. C. (1998). Effect of feeding aflatoxin B1 on feed consumption through naturally contaminated feeds, *Indian J. Anim. Sci.*, 400–401.

Čolović, R., Puvača, N., Cheli, F., Avantaggiato, G., Greco, D., Đuragić, O., Kos, J. and Pinotti, L. (2019). Decontamination of mycotoxin-contaminated feedstuffs and compound feed, *Toxins* 11(11), 617–634.

Colvin, B. M., Cooley, A. J. and Beaver, R. W. (1993). Fumonisin toxicosis in swine: clinical and pathologic findings, *J. Vet. Diagn. Invest.* 5(2), 232–241.

Cook, W. O., Richard, J. L., Osweiller, G. D. and Trampel, D. W. (1986). Clinical and pathologic changes in acute bovine aflatoxicosis: rumen motility and tissue and fluid concentrations of aflatoxins B1 and M1, *Am. J. Vet. Res.* 47(8), 1817–1825.

Commission Directive (EC) 100/2003. Available at: https://eur-lex.europa.eu/eli/dir/2003/100/oj.

Commission Recommendation (EC) 576/2006. Available at: https://eur-lex.europa.eu/eli/reco/2006/576/oj.

Compliance Policy Guide (2019). Action Levels for Aflatoxins in Animal Food. 683.100 FDA, FDA-2013-S-0610. Available at https://www.fda.gov/regulatory-information/search-fda-guidance-documents/cpg-sec-683100-action-levels-aflatoxins-animal-feeds.

Cortinovis, C., Caloni, F., Schreiber, N. B. and Spicer, L. J. (2014). Effects of fumonisin B1 alone and combined with deoxynivalenol or zearalenone on porcine granulosa cell proliferation and steroid production, *Theriogenology* 81(8), 1042–1049.

Dai, M., Jiang, S., Yuan, X., Yang, W., Yang, Z. and Huang, L. (2016). Effects of zearalenone-diet on expression of ghrelin and PCNA genes in ovaries of post-weaning piglets, *Anim. Reprod. Sci.* 168, 126–137.

Deepa, N. and Sreenivasa, M. Y. (2017). Fumonisins: a review on its global occurrence, epidemiology, toxicity and detection, *J. Vet. Med. Res.* 4, 1093–1103.

Dell'Orto, V., Baldi, G. and Cheli, F. (2015). Mycotoxins in silage: checkpoints for effective management and control, *World Mycotoxin J.* 8(5), 603–617.

Denli, M. and Perez, J. F. (2010). Ochratoxins in feed, a risk for animal and human health: control strategies, *Toxins* 2(5), 1065–1077.

Driehuis, F., Spanjer, M. C., Scholten, J. M. and Te Giffel, M. C. (2008). Occurrence of mycotoxins in feedstuffs of dairy cows and estimation of total dietary intakes, *J. Dairy Sci.* 91(11), 4261–4271.

Duarte, S. C., Pena, A. and Lino, C. M. (2010). A review on ochratoxin A occurrence and effects of processing of cereal and cereal derived food products, *Food Microbiol.* 27(2), 187–198.

EFSA (European Food Safety Authority) (2004). *Opinion of the Scientific Panel on Contaminants in the Food Chain on a Request from the Commission Related to Aflatoxin B1 as Undesirable Substance in Animal Feed*. European Food Safety Authority (EFSA), EFSA-Q-2003-035.

EFSA (2017a). Risks for animal health related to the presence of zearalenone and its modified forms in feed, *EFSA J.* 15(7), e04851.

EFSA (2017b). Risks to human and animal health related to the presence of deoxynivalenol and its acetylated and modified forms in food and feed, *EFSA J.* 15(9), e04718.

Egbuta, M. A., Mwanza, M., Phoku, J. Z., Chilaka, C. A. and Dutton, M. F. (2016). Comparative analysis of mycotoxigenic fungi and mycotoxins contaminating soya bean seeds and processed soya bean from Nigerian markets, *Adv. Microbiol.* 6, 1130–1139.

Ferrao, J. Bell, V., Chabite, I. T. and Fernandes, T. H. (2017). Mycotoxins, food and health, *J. Nutr. Health Food Sci.* 5(7), 1–10.

Fink-Gremmels, J. (2008a). Mycotoxins in cattle feeds and carry-over to dairy milk: a review, *Food Addit. Contam. Part A* 25(2), 172–180.

Fink-Gremmels, J. (2008b). The role of mycotoxins in the health and performance of dairy cows, *Vet. J.* 176(1), 84–92.

Food and Agriculture Organization of the United Nations (FAO) (2007). *Training Manual On-Farm Mycotoxin Control in Food ad Feed Grain*. Rome: FAO.

Foroud, N. A. and Eudes, F. (2009). Trichothecenes in cereal grains, *Int. J. Mol. Sci.* 10(1), 147–173.

Garlich, J. D., Tung, H. T. and Hamilton, P. B. (1973). Effects of short-term feeding of aflatoxin on egg production and some plasma constituents of laying hen, *Poult. Sci.* 52(6), 2206-2211.
Gazzotti, T., Lugoboni, B., Zironi, E., Barbarossa, A., Serraino, A. and Pagliuca, G. (2009). Determination of fumonisin B1 in bovine milk by LC-MS/MS, *Food Control* 20(12), 1171-1174.
Gerez, J. R., Desto, S. S., Frederico, A. P. and Bracarense, R. L. (2017). Deoxynivalenol induces toxic effects in the ovaries of pigs: an ex vivo approach, *Theriogenology* 90, 94-100.
Glavitis, R. and Vanyi, A. (1995). More important mycotoxicosis in pigs, *Magy. Allatorvosak Lapja* 50, 407-420.
Gruber-Dorninger, C., Jenkins, T. and Schatzmayr, G. (2019). Global mycotoxin occurrence in feed: a ten-year survey, *Toxins* 11(7), 375-399.
Harvey, R. B., Huff, W. E., Kubena, L. F., Corrier, D. E. and Phillips, T. D. (1988). Progression of aflatoxicosis in growing barrows, *Am. J. Vet. Res.* 49(4), 482-487.
Helferich, W. G., Garrett, W. N., Hsieh, D. P. H. and Baldwin, R. L. (1986). Feedlot performance and tissue residues of cattle consuming diets containing aflatoxins, *J. Anim. Sci.* 62(3), 691-696.
Heperkan, D., Güler, F. K. and Oktay, H. I. (2012). Mycoflora and natural occurrence of aflatoxin, cyclopiazonic acid, fumonisin and ochratoxin A in dried figs, *Food Addit. Contam. Part A* 29(2), 277-286.
Hesseltine, C. W. (1974). Natural occurrece of mycotoxins in cereals, *Mycopathol. Mycol. Appl.* 53(1), 141-153.
Howarth, B. and Wyatt, R. D. (1976). Effect of dietary aflatoxin on fertility, hatchability and progeny performance of broiler breeder hens, *Appl. Environ. Microbiol.* 31(5), 680-684.
Huff, W. E., Kubena, L. F., Harvey, R. B. and Doerr, J. A. (1988). Mycotoxin interactions in poultry and swine, *J. Anim. Sci.* 66(9), 2351-2355.
Hussein, H. S. and Brasel, J. M. (2001). Toxicity, metabolism, and impact of mycotoxins on humans and animals, *Toxicology* 167(2), 101-134.
Iheshiulor, O. O. M., Esonu, B. O., Chuwuka, O. K., Omede, A. A., Okoli, I. C. and Ogbuewu, I. P. (2011). Effects of mycotoxins in animal nutrition: a review, *Asian J. Anim. Sci.* 5(1), 19-33.
Jorgensen, K. (1998). Survey of pork, poultry, coffee, beer and pulses for ochratoxin A, *Food Addit. Contam.* 15(5), 550- 554.
Kamle, M., Mahato, D. K., Devi, S., Lee, K. E., Kang, S. G. and Kumar, P. (2019). Fumonisins: impact on agriculture, food, and human health and their management strategies, *Toxins* 11(6), 328-350.
Kew, M. C. (2013). Aflatoxins as a cause of hepatocellular carcinoma, *J. Gastrointestin. Liver Dis.* 22(3), 305-310.
Khaneghah, A. M., Fakhri, Y., Gahruie, H. H., Niakousari, M. and Sant'Ana, A. S. (2019). Mycotoxins in cereal-based products during 24 years (1983-2017): a global sysrematic review, *Trends Food Sci. Technol.* 91, 95-105.
Korosteleva, S. N., Smith, T. K. and Boermans, H. J. (2009). Effects of feed naturally contaminated with Fusarium mycotoxins on metabolism and immunity of dairy cows, *J. Dairy Sci.* 92(4), 1585-1593.
Kőszegi, T. and Poór, M. (2016). Ochratoxin A: molecular interactions, mechanisms of toxicity and prevention at the molecular level, *Toxins* 8(4), 111.

Krogh, P. (1991). Porcine nephropathy associated with ochratoxin A. In: Smith, J. E. and Anderson, R. A. (Eds), *Mycotoxins and Animal Foods*. Boca Raton, FL: CRC Press, pp. 627–645.

Kuiper-Goodman, T. and Scott, P. M. (1989). Risk assessment of the mycotoxin ochratoxin A, *Biomed. Environ. Sci.* 2(3), 179–248.

Langevin, F., Eudes, F. and Comeau, A. (2004). Effect of trichothecenes produced by Fusarium graminearum during Fusarium head blight development in six cereal species, *Eur. J. Plant Pathol.* 110(7), 735–746.

Ledoux, D. R., Brown, T. P., Weibking, T. S. and Rottinghaus, G. E. (1992). Fumonisin toxicity in broiler chicks, *J. Vet. Diagn. Invest.* 4(3), 330–333.

Lee, H. J. and Ryu, D. (2017). Worldwide occurrence of mycotoxins in cereals and cereal-derived food products: public health perspectives of their co-occurrence, *J. Agric. Food Chem.* 65(33), 7034–7051.

Levasseur-Garcia, C. (2018). Updated overview of infrared spectroscopy methods for detecting mycotoxins on cereals (corn, wheat, and barley), *Toxins* 10(1), 38–50.

Liew, W. P. P. and Mohd-Redzwan, M. (2018). Mycotoxin: its impact on gut health and microbiota, *Front. Cell. Infect. Microbiol.* 8, 1–17.

Liu, J. and Applegate, T. (2020). Zearalenone (ZEN) in livestock and poultry: dose, toxicokinetics, toxicity and estrogenicity, *Toxins* 12(6), 377–397.

Magan, N., Medina, A. and Aldred, D. (2011). Possible climate change effects on mycotoxin contamination of food crops pre- and post-harvest, *Plant Pathol.* 60, 150–163.

Marchese, S., Polo, A., Ariano, A., Velotto, S., Costantini, S. and Severino, L. (2018). Aflatoxin B1 and M1: biological properties and their involvement in cancer development, *Toxins* 10(6), 214–232.

Medina, A., Rodriguez, A. and Magan, N. (2014). Effect of climate change on Aspergillus flavus and aflatoxin b1 production, *Front. Microbiol.* 5, 348.

Meissonnier, G. M., Bracarense, A. P., Bertin, G., Galtier, P. and Oswald, I. P. (2008). Toxicity of type-A trichothecenes in farm animals. In: Oswald, I. P. and Taranu, I. (Eds) *Mycotoxins in Farm Animals*. Kerala, India: Research Sign Post, pp. 131–154.

Minervini, F. and Dell'Aquila, M. E. (2008). Zearalenone and reproductive function in farm animals, *Int. J. Mol. Sci.* 9(12), 2570–2584.

Misihairabgwi, J. M., Ezekiel, C. N., Sulyok, M., Shephard, G. S. and Krska, R. (2019)Mycotoxin contamination of foods in Southern Africa: a 10-year review (2007–2016), *Crit. Rev. Food Sci. Nutr.* 59(1), 43–58.

Mupunga, I., Mngqawa, P. and Katerere, D. R. (2017). Peanuts, aflatoxins and undernutrition in children in sub-Saharan Africa, *Nutrients* 9(12), 1287–1298.

Murugesan, G. R., Ledoux, D. R., Naehrer, K., Berthiller, F., Applegate, T. J., Grenier, B., Phillips, T. D. and Schatzmayr, G. (2015). Prevalence and effects of mycotoxins on poultry health and performance, and recend development in mycotoxin counteracting strategies, *Poult. Sci.* 94(6), 1298–1315.

Neng, J., Zhang, Q. and Sun, P. (2020). Application of surface-enhanced Raman spectroscopy in fast detection of toxic and harmful substances in food, *Biosens. Bioelectron.* 167, 112480–112489.

Olsson, J., Borjesson, T., Lundstedt, T. and Schnuerer, J. (2002). Detection and quantification of ochratoxin and deoxinivalenol in barley grain by GC-MS and electronic nose, *Int. J. Food Microbiol.* 72(3), 203–214.

Ostry, V., Malir, F., Toman, J. and Grosse, Y. (2017). Mycotoxins as human carcinogens-the IARC Monographs classification, *Mycotoxin Res.* 33(1), 65–73.

Ottoboni, M., Pinotti, L., Tretola, M., Giromini, C., Fusi, E., Rebucci, R., Grillo, M., Tassoni, L., Foresta, S., Gastaldello, S., Furlan, V., Maran, C., Dell'Orto, V. and Cheli, F. (2018). Combining e-nose and lateral flow immunoassays (LFIAs) for rapid occurrence/co-occurrence aflatoxin and fumonisin detection in maize, *Toxins* 10(10), 416–426.

Parry, D. W., Jenkinson, P. and McLeod, L. (1995). Fusarium ear blight (scab) in small grain cereals–a review, *Plant Pathol.* 44(2), 207–238.

Patterson, D. S. P., Shreeve, B. J., Roberts, B. A., Berrett, S., Brush, P. J., Glancy, E. M. and Krogh, P. (1981). Effect on calves of barley naturally contaminated with ochratoxin A and groundnut meal contaminated with low concentration of aflatoxin b1, *Res. Vet. Sci.* 31(2), 213–218.

Paul, P. S., Johnson, D. W., Mirocha, C. J., Soper, F. F., Thoen, C. O., Muscoplat, C. C. and Weber, A. F. (1977). In vitro stimulation of bovine peripheral blood lymphocytes: suppression of phytomitogen and specific antigen lymphocyte responses by aflatoxin, *Am. J. Vet. Res.* 38(12), 2033–2035.

Peraica, M., Domijan, A. M., Jurjevic, Z. and Cvjetkovic, B. (2002). Prevention of exposure to mycotoxins from food and feed, *Arh. Hig. Rada Toksikol.* 53(3), 229–237.

Pereira, C. S., Cunha, S. C. and Fernandes, J. O. (2019). Prevalent mycotoxins in animal feed: occurrence and analytical methods, *Toxins* 11(5), 290–351.

Pestka, J. J. (2007). Deoxynivalenol: toxicity, mechanisms and animal health risks, *Anim. Feed Sci. Technol.* 137(3–4), 283–298.

Pierron, A., Alassane-Kpembi, I. and Oswald, I. P. (2016). Impact of mycotoxin on immune response and consequences for pig health, *Anim. Nutr.* 2(2), 63–68.

Pinotti, L., Ottoboni, M., Giromini, C., Dell'Orto, V. and Cheli, F. (2016). Mycotoxin contamination in the EU feed supply chain: a focus on cereal byproducts, *Toxins* 8(2), 45.

Piotrowska, M., Slizewska, K. and Biernasiak, J. (2013). Mycotoxin in cereal and soybean-based food and feed. In: El-Shemy, H. A. (Ed), *Soybean-Pest Resistance*. Giza, Egypt: Cairo University, pp. 3565–3782.

Pitt, J. I. (2000). Toxigenic fungi: which are important?, *Med. Mycol.* 38 (Suppl. 1), 17–22.

Pizzo, F., Caloni, F., Schutz, L. F., Totty, M. L. and Spicer, L. J. (2015). Individual and combined effects of deoxynivalenol and a-zearalenol on cell proliferation and steroidogenesis of granulosa cells in cattle, *Environ. Toxicol. Pharmacol.* 40(3), 722–728.

Pizzo, F., Caloni, F., Schreiber, N. B., Cortinovis, C. and Spicer, L. J. (2016). In vitro effects of deoxynivalenol and zearalenone major metabolites alone and combined, on cell proliferation, steroid production and gene expression in bovine small-follicle granulosa cells, *Toxicon* 109, 70–83.

Rai, A., Das, M. and Tripathi, A. (2019). Occurrence and toxicity of a fusarium mycotoxin, zearalenone, *Crit. Rev. Food Sci. Nutr.* 60(16), 2710–2729.

Ranzenigo, G., Caloni, F., Cremonesi, F., Aad, P. Y. and Spicer, L. J. (2008). Effects of Fusarium mycotoxins on steroid production by porcine granulosa cells, *Anim. Reprod. Sci.* 107(1–2), 115–130.

Reddy, L. and Bhoola, K. (2010). Ochratoxins–food contaminants: impact on human health, *Toxins* 2(4), 771–779.

Ross, P. F., Rice, L. G., Reagor, J. C., Osweiler, G. D., Wilson, T. M., Nelson, H. A., Owens, D. L., Plattner, R. D., Harlin, K. A. and Richard, J. L., Colvin, B. M. and Banton, M. I. (1991). Fumonisin B1 concentrations in feeds from 45 confirmed equine leukoencephalomalacia cases, *J. Vet. Diagn. Invest.* 3(3), 238–241.

Scott, P. M. (2012). Recent research on fumonisins: a review, *Food Addit. Contam. Part A* 29(2), 242–248.

Seefelder, W., Gossmann, M. and Humpf, H. U. (2002). Analysis of fumonisin B1 in Fusarium proliferatum-infected asparagus spears and garlic bulbs from Germany by liquid chromatography - electrospray ionization mass spectrometry, *J. Agric. Food Chem.* 50(10), 2778–2781.

Shotwell, O. L., Hesseltiney, C. W. and Goulden, M. L. (1969). Ochratoxin A occurrence as natural contaminant of a corn sample, *Appl. Microbiol.* 17(5), 765–766.

Smith, M. C., Madec, S., Coton, E. and Hymery, N. (2016). Natural Co-occurrence of mycotoxins in foods and feeds and their in vitro combined toxicological effects, *Toxins* 8(4), 94.

Southern, L. L. and Clawson, A. J. (1979). Effects of aflatoxins on finishing swine, *J. Anim. Sci.* 49(4), 1006–1011.

Stępień, L., Koczyk, G. and Waśkiewicz, A. (2011). Genetic and phenotypic variation of Fusarium proliferatum isolates from different host species, *J. Appl. Genet.* 52(4), 487–496.

Streit, E., Naehrer, K., Rodrigues, I. and Schatzmayr, G. (2013). Mycotoxin occurrence in feed and feed raw materials worldwide: long-term analysis with special focus on Europe and Asia, *J. Sci. Food Agric.* 93(12), 2892–2899.

Sun, X. D., Su, P. and Shan, H. (2017). Mycotoxin contamination of maize in China, *Compr. Rev. Food Sci. Food Saf.* 16(5), 835–849.

Sypecka, Z., Kelly, M. and Brereton, P. (2004). Deoxynivalenol and zearalenone residues in eggs of laying hens fed with a naturally contaminated diet: effects on egg production and estimation of transmission rates from feed to eggs, *J. Agric. Food Chem.* 52(17), 5463–5471.

Thiel, P. G., Shephard, G. S., Sydenham, E. W., Marasas, W. F. O., Nelson, P. E. and Wilson, T. M. (1991). Levels of fumonisins B1 and B2 in feeds associated with confirmed cases of equine leukoencephalomalacia, *J. Agric. Food Chem.* 39(1), 109–111.

Tobias, S., Rajic, I. and Vanyi, A. (1992). Effect of T-2 toxin on egg production and hatchability in laying hens, *Acta Vet. Hung.* 40(1-2), 47–54.

U.S. Food and Drug Administraton (FDA). Available at: https://www.fda.gov/animal-veterinary/biological-chemical-and-physical-contaminants-animal-food/chemical-hazards.

Vendruscolo, C. P., Frias, N. C., de Carvalho, C. B., de Sá, L. R. M., Belli, C. B. and Baccarin, R. Y. A. (2016). Leukoencephalomalacia outbreak in horses due to consumption of contaminated hay, *J. Vet. Intern. Med.* 30(6), 1879–1881.

Vesonder, R., Haliburton, J., Stubblefield, R., Gilmore, W. and Peterson, S. (1991). Aspergillus flavus and aflatoxins B1, B2, and M1 in corn associated with equine death, *Arch. Environ. Contam. Toxicol.* 20(1), 151–153.

World Health Organization (1993). Evaluation of carcinogenic risks to humans 56, 1–599.

Wyatt, R. D., Doerr, J. A., Hamilton, P. B. and Burmeister, H. R. (1975). Egg production, shell thickness, and other physiological parameters of laying hens affected by T-2 toxin, *Appl. Microbiol.* 29(5), 641–645.

Zachariasova, M., Dzuman, Z., Veprikova, Z., Hajkova, K., Jiru, M., Vaclavikova, M., Zachariasova, A., Pospichalova, M., Florian, M. and Hajslova, J. (2014). Occurrence of multiple mycotoxins in European feedingstuffs, assessment of dietary intake by farm animals, *Anim. Feed Sci. Technol.* 193, 124–140.

Zheng, W., Wang, B., Li, X., Wang, T., Zou, H., Gu, J., Yuan, Y., Liu, X., Bai, J., Bian, J. and Liu, Z. (2018). Zearalenone promotes cell proliferation or causes cell death?, *Toxins* 10(5), 184–200.

Zimmerli, B. and Dick, R. (1996). Ochratoxin A in table wine and grape-juice: occurrence and risk assessment, *Food Addit. Contam.* 13(6), 655–668.

Zinedine, A., Soriano, J. M., Juan, C., Mojemmi, B., Moltó, J. C., Bouclouze, A., Cherrah, Y., Idrissi, L., El Aouad, R. and Mañes, J. (2007). Incidence of ochratoxin A in rice and dried fruits from Rabat and Salé area, Morocco, *Food Addit. Contam.* 24(3), 285–291.

Chapter 7

Risk management systems for prevention and control of contaminants in animal feed

Regiane R. Santos, Schothorst Feed Research, The Netherlands

1 Introduction

It is estimated that world compound feed production is approaching 1 billion tonnes per year (FAO 2020). To achieve optimal animal production, feed design systems need to combine the optimum level of nutrients and feed additives together with good hygienic management. Feed design needs to optimize the supply of nutrients but minimize excretion of excess nutrients that may lead to environmental pollution, for example, of phosphorus and nitrogen. Feed safety contributes to these goals as well as being essential to animal health and welfare. It is also a precondition for food safety and human health. Legislation in regions such as the EU requires that animals that enter the food chain may be fed only with safe feed (Commission Regulation (EU) No. 68/2013 on the Catalogue of feed materials; Regulation EC 767/2009).

Several feed contaminants of biological, chemical or physical origin are of concern in animal feed manufacture, because they can enter the feed chain at any stage of feed production, from raw material production through processing to diet formulation. Ensuring feed safety requires systematic prevention and control systems. Each hazard must first be identified and characterized. Subsequently, the potential exposure risk must be assessed

http://dx.doi.org/10.19103/AS.2020.0083.14

and the risk characterization then determined. Depending on the hazard, management tools are developed to protect feed and feed ingredients from contamination during each stage of production, whether in the field, handling, storage, transport or processing. Control systems used by governments, feed producers and market chains include good agricultural practices (GAPs), good manufacturing practices (GMPs) and/or hazard analysis and critical control point (HACCP) systems.

National, regional and international authorities are responsible for providing risk assessments together with maximum tolerable limits for contaminants. The main international and regional authorities are the Food and Agricultural Organization of the United Nations (FAO), World Health Organization (WHO), the Codex Alimentarius Commission (CAC), the US Food and Drug Administration (FDA) and the European Food Safety Authority (EFSA). Regulatory and surveillance agencies such as the US Department of Agriculture (USDA) Animal and Plant Health Inspection Services (APHIS), the USDA Agriculture Research Service (ARS) and the USDA Food Safety and Inspection Service (FSIS), the Canadian Food Inspection Agency (CFIA), Food Standards Agency (FSA) in UK, the State Food and Drug Administration of China (SFDA), the Food Safety and Standards Authority of India (FSSAI) and the Ministry of Health, Labor and Welfare (MHLW) in Japan have been producing relevant data and guidelines to safeguard food safety (Pandey et al. 2020). However, these systems are not yet applied or harmonized globally, given different approaches by regional authorities in different countries and, in some countries, the absence of transparent information on risk assessment. Such uniformity is needed because of globalization, modern trading, the prospects of utilizing alternative feedstuffs and the risks of foodborne zoonotic disease outbreaks. It is also important to consider potential interactions between contaminants and not just guidelines per contaminant or contaminant group. The purpose of this chapter is to summarize the main risk management systems to prevent and control contaminants in animal feed.

2 Contaminants in animal feed

Livestock and humans often compete for the same feedstuffs (e.g. cereals are used for both food and feed). Commodities of the highest quality, including the safest contaminant levels, are intended for human consumption. The remainder is used for animal feed or other purposes such as biofuel. Contaminants encountered in animal feed include bacteria, viruses, prions, parasites, metals, pesticides, molds, toxins and allergens. This chapter is not intended to provide a detailed description of feed contaminants or their residuals in the food chain (e.g. veterinary drugs), but it is necessary to give a brief overview of the risks posed by contaminants in animal feed. For details on the main contaminants found in animal feed, as well as regulations of maximum levels accepted by the EU and the United States, see Gerssen et al. (2019) and Verstraete (2013), respectively.

The main contaminating pathogens are *Salmonella* and the prion protein causing bovine spongiform encephalopathy (BSE). Both agents are of particular importance because they can enter the human food chain and cause potentially fatal disease. *Salmonella* can be eliminated by thermal treatment during pelleting and extrusion; chemical treatments are also used in mash diets with variable rates of success (Wales et al. 2010). The prion responsible for BSE can be found in feed based on processed animal protein; when BSE contamination is confirmed, the raw material is prohibited for use in animal feed. In the United States meat and bone meal from cattle can only be fed to non-ruminants whilst meat and bone meal from non-ruminants can be fed to cattle (CMPAF; U.S. Food and Drug Administration (FDA). Prohibited material includes the brain and spinal cords from cattle of 30 months of age and older (for a review, see Pandey et al. 2020). Recent studies have also identified potential risks of virus transmission via feed, though more research is needed to confirm this risk. Porcine epidemic diarrhea (PED) and porcine delta coronavirus (PDCoV), both swine enteric coronaviruses, together with the risk of introducing African swine fever (ASF), have highlighted the possibility of viral disease transmission via feed ingredient contamination (Gordon et al. 2019).

Contamination risks from metals and pesticides are also carefully controlled in food for human consumption and regulated by strict legislation. Robust regulation is also applied in animal feed due to the risk of tissue deposition and food chain contamination. However, exposure characterization and risk assessment must be based on realistic scenarios to avoid overestimation of risk resulting in economic loss or underestimation of risk resulting in health risks for consumers. To achieve this, a transfer factor, that is, the transfer of chemical compounds from animal feed (mg/kg on a dry weight basis) to animal products (in mg/kg on a wet weight basis), has been suggested (Leeman et al. 2007). Although such a system is more accurate than simple estimation based on the worst-case scenario, many variables must be included such as time of exposure, animal age, gender and interaction with other compounds or pathogens (the most difficult parameter to account for). Not all the metals present in animal diets are essential to maintain physiological functions; As, Cd, F, Pb and Hg are considered feed contaminants which can enter the food chain through animal consumption (Reg. 2002/32/EC). In the case of pesticides, a risk assessment must consider a range of factors in addition to dietary intake, including management, type and quality of raw materials, additives, soil ingestion and accidental contamination (Hejna et al. 2018).

Mycotoxins are common contaminants in animal feed. Mycotoxins are undesirable natural toxins present in almost all feedstuffs which are usually contaminated by a mixture of these chemicals. Simply preparing diets following recommended levels might avoid acute toxicity, but will not be a guarantee of optimal animal performance because the guidelines were developed based on contamination by individual mycotoxins and not on the interaction of mycotoxins. The European Union is running a project called MYCHIF (**Myc**otoxin mixtures in food and feed: **H**olistic, **I**nnovative, **F**lexible) on risk assessment modeling to understand the effects of contamination of feedstuffs with multiple mycotoxins (Battilani et al. 2020). A bottleneck is the scarce number of studies performed with naturally contaminated feed, the varied distribution of mycotoxins in different dietary matrices around the world, and the fact that most of the data related to mycotoxin interaction effects on animal health were performed with unrealistically high levels of two or three mycotoxins. It is also necessary to evaluate transgenerational toxicity (Schoevers et al. 2012), as well as the interaction of mycotoxins with host-microbiota products (Silva et al. 2019). Carry-over of these contaminants into the food chain will depend on bioaccumulation and toxicokinetics that may vary within animal species and type of food of animal origin, that is, milk, eggs and meat (Cheli et al. 2013).

Last but not least, biogenic amines are among key allergens that may contaminate animal feed. Biogenic amines are polyamines synthesized due to amino acid decarboxylation and are usually found in by-products that have suffered from microbial spoilage, such as bone meal, blood meal, poultry meal, feather meal and, in particular, fish meal. Histamine is the most studied biogenic amine due to the high risk to human health (Morrow et al. 1991). These allergens can also affect animal health. Although these substances are resistant to pelleting and other processing techniques, contamination levels can be controlled by proper hygiene measures during preparation and storage. It is important to bear in mind that the production of biogenic amines is usually accompanied by contamination of meal by microorganisms, that is, in feed material which is already of poor quality (Jaw et al. 2012). Consumption of animal products with biogenic amines is a particular risk to human health since processing, canning or cooking does not destroy these allergens (Ruiz-Capillas and Herrero 2019).

3 New and alternative feed ingredients: safety issues

A sustainable world requires the prevention of waste, reuse and recycling of material and energy recovery up to disposal (Directive 2008/98/EC; van Raamsdonk et al. 2019). The circular economy will play an important role in agriculture, as shown by a program in this area initiated by the European Commission in 2015. Feed is a key stage at the beginning of the food chain.

The safety of new feed materials must be considered before entering the feed chain.

The edible portion of slaughtered animals suitable for human consumption is 68%, 62% and 54% for poultry, swine and cattle, respectively (Alm 2012). This means that between 32% and 46% of the weight of food-producing animals is surplus. According to EU regulations not all this surplus can be recycled to produce livestock feed. These residues are first categorized into three risk categories (Table 1) and a rendering process is mandatory before the material can enter the feed chain. When animal by-products of different categories are mixed, EU regulations require that the mixture be handled in accordance with the standards laid down for the by-product belonging to the highest risk category.

Rendering of animal carcasses is a common procedure to obtain nutrient sources for animal consumption, for example, meat meal, feather meal, bone meal and fish meal. This process takes place at high temperatures which will usually destroy pathogens. However, this form of disposal is not applied to animal parts contaminated with the bovine spongiform encephalopathy (BSE) prion agent. This should be considered specified risk material (SRM) where rendering is not applicable (Table 1).

Edible by-products can be used in animal feed. Although not classified as processed animal proteins (PAP), milk and milk products, eggs and egg products, collagen and gelatine are authorized in animal feed. Since 2013 swine and poultry PAP can be used as feed in the aquaculture sector in the EU, and fishmeal can be used to feed pigs, poultry, horses and farmed fish. Given environmental concerns, there has been a discussion of European regulation of processed animal proteins, related particularly to the use of soya as a protein source due to Europe's deficit of protein-rich feed ingredients (Jedrerek et al. 2016). Discussions about using by-products must balance safety and environmental concerns, which could be achieved by harmonized legislation and robust monitoring systems (van Raamsdonk et al. 2019).

The use of insects as an alternative source of proteins is controversial since they are considered an animal protein and their impact on feed efficiency in farm animals is not always successful (van Raamsdonk et al. 2017). The presence of contaminants like pathogens, mycotoxins, heavy metals, drug and pesticide residues, as well as toxins produced by the insect itself, should be considered. Although levels of contaminants in insect larvae are usually below the recommended maximum (EC, WHO and Codex) levels (Charlton et al. 2015), chronic exposure may lead to bioaccumulation of contaminants to significant levels. A recent review (Schrögel and Wätjen 2019) found mycotoxins and heavy metals to be the main contaminants in insects. Although levels were not serious, a starving period is recommended before feeding animals.

Table 1 Management of three categories of animal by-products, according to EU regulation 1069/2009

Categories		Disposal and use (EU legislation)
1:	• Entire bodies and body parts of animals infected of suspected of being infected by a TSE, as well those killed in the context of TSE eradication measures; • Animals other than farmed and wild animals; • Animals used for experiments (Directive 86/609/EEC); • Wild animals, when suspected of being infected with diseases communicable to humans or animals; • Animal by-products derived from animals submitted to illegal treatment as defined in Directive 96/22/EC; • Animal by-products containing residues of other substances and environmental listed in Directive 96/23/EC; • Animal by-products collected during the treatment of waste water; • Catering waste from means of transport operating internationally; • Mixtures of Category 1 material with either Category 2 material or Category 3 material or both.	Incineration in an approved plant; Bury in an authorized landfill; Used as a fuel for combustion with or without prior processing.
2	• Manure, non-mineralised guano and digestive tract content; • Animal by-products collected during the treatment of waste water from establishments or plants processing Category 2 material or from slaughterhouse other than those covered by Article 8; • Animal by-products containing residues of authorised substances or contaminants exceeding the permitted levels as referred to in Article 15(3) of Directive 96/23/EC; • Products of animal origin which have been declared unfit for human consumption due to the presence of foreign bodies in those products;	Incineration and/or rendering, or at an authorized landfill site; Recycling for the manufacturing of organic fertilisers or soil improvers to be placed on the market; Composted or transformed into biogas; Applied to land without processing, in the case of manure, digestive tract content separated from the digestive tract, milk, milk-based products and colostrum which the competent authority does not consider to present a risk for the spread of any serious transmissible disease; Used as a fuel for combustion with or without prior processing.

	• Products of animal origin, other than Category 1 material, that are imported or introduced from a third country and fail to comply with Community veterinary legislation, or dispatched to another Member State and fail to comply with requirements laid down or authorised by Community legislation; • Animals and parts of animals, that died other than by being slaughtered for human consumption, foetuses, and dead-in-shell poultry; • Mixtures of Category 2 material with Category 3 material; • Animal by-products other than Category 1 material or Category 3 material.	
3	• Carcasses and parts of animals slaughtered or bodies or parts of animals killed, and which are fit for human consumption in accordance with Community legislation, but are not intended for human consumption for commercial reasons; • Heads of poultry, hides and skins, including trimmings and splitting thereof, horns and feet, including the phalanges and the carpus and metacarpus bones, tarsus and metatarsus bones of animals, other than ruminants requiring TSE testing, and ruminants which have been tested with a negative result, pig bristles, feathers; • Animal by-products from poultry and lagomorphs slaughtered on the farm, which did not show any signs of disease communicable to humans or animals; • Blood of animals which did not show any signs of disease communicable through blood to humans or animals obtained from animals other than ruminants requiring TSE testing, and ruminants which have been tested with a negative result;	Incineration, with or without prior processing; Recovered or disposed of by co-incineration, with or without prior processing, if the Category 3 material is waste; Disposed of in an authorised landfill, following processing; Processed, except in the case of Category 3 material which has changed through decomposition or spoilage so as to present an unacceptable risk to public or animal health; Production of raw pet food, to be placed on the market in accordance with Article 35; Composted or transformed into biogas; Used as a fuel for combustion with or without prior processing; Used for the manufacture of derived products referred to in Articles 33, 34 and 36 and placed on the market in accordance with those Articles.

(Continued)

Table 1 (*Continued*)

Categories	Disposal and use (EU legislation)
• Animal by-products arising from the production of products intended for human consumption, including degreased bones, greaves and centrifuge or separator sludge from milk processing; • Products of animal origin, foodstuffs, pet foods, or feedstuffs containing products of animal origin, which are no longer intended for human consumption for commercial reasons or due to problems of manufacturing or packaging defects or other defects from which no risk to public or animal health arise; • Blood, placenta, wool, feathers, hair, horns, hoof cuts and raw milk originating from live animals that did not show any signs of disease communicable through that product to humans or animals; • Aquatic animals, and parts of such animals, except sea mammals, which did not show any signs of disease communicable to humans or animals; • Animal by-products from aquatic animals originating from establishments or plants manufacturing products for human consumption; • Material originating from animals which did not show any signs of disease communicable through that material to humans or animals: shells from shellfish with soft tissue or flesh, hatchery by-products, eggs, egg by-products, including egg shells, day-old chicks killed for commercial reasons, aquatic and terrestrial invertebrates other than species pathogenic to humans or animals; • Animals and parts thereof of the zoological orders of Rodentia and Lagomorpha, except Category 1; • Hides and skins, hooves, feathers, wool, horns, hair and fur originating from dead animals that did not show any signs of disease; • Adipose tissue from animals which did not show any signs of disease communicable through that material to humans or animals; • Catering waste other than as referred to in Article 8(f).	

TSE: Transmissible spongiform encephalopathies.

It has been shown that, when fed feedstuffs contaminated with high levels of the aflatoxin B1 mycotoxin, black soldier fly larvae (*Hermetia illucens*, BSF) and yellow mealworm (*Tenebrio molitor*, YMW) can help decontaminate feed materials (Bosch et al. 2017). Accumulation of heavy metals may vary in accordance with the heavy metal, insect species and developmental stage. Specified insect species can be added to poultry diets (EU and Canada) and used in aquaculture (the United States and Canada). Legislation varies between countries. Although the use of insects is allowed in the United States and Canada, its use in fish diets was recently banned in Europe due to the risk of prion-derived diseases (BfR 2019). There are no specific laws on the regulation of insect for animal feed in China and South Korea, while its use is forbidden in North Korea, where insects are considered animal-based protein (Sogari et al. 2019). In Australia and Mexico, feed materials like insects usually do not require specific legislation (Lähteenmäki-Uutela et al. 2017). A summary of legislation on the use of insects in feed, as well as the main list of insect species allowed in Europe and North America, is provided in Table 2 and is summarized in IPIFF (2019) and Sogari et al. (2019).

Table 2 List of insect species allowed to be included in animal diets

	Europe	USA	Canada
Allowed insect species	Black soldier fly (*Hermetia illucens*)	Black soldier fly (*Hermetia illucens*)	Black soldier fly (*Hermetia illucens*)
	Common house fly (*Musca domestica*)		
	House cricket (*Acheta domesticus*)		
	Banded cricket (*Gryllodes sigillatus*)		
	Field cricket (*Gryllus assimilis*)		
	Yellow mealworm (*Tenebrio molitor*)		
	Lesser mealworm (*Alphitobius diaperinus*)		
Authority	EFSA	FDA	CFIA
Regulation	EU decisions/regulations	FFDCA	FAFR
Definition of insect as feed	Processed animal proteins (PAPs) Authorized fat from insects in feed	Feed additive; approval list or GRAS is needed	Feed raw material; each authorization is evaluated separately

EFSA: European Food Safety Authority; FDA: Federal Food and Drug Administration; FFDCA: Federal Food, Drug, and Cosmetic Act; GRAS: Generally Recognized as Safe; CFIA: Canadian Food Inspection Agency; FAFR: Food Act and Feeds Regulation.

Practices relating to rearing, handling, processing and storage of insects need to guarantee safety and traceability (Verbeek et al. 2015). The International Platform of Insects for Food and Feed (IPIFF 2019) has issued recommendations on good practices for insect production governing safety procedures and HACCP principles. When rearing insects in the EU, producers must use legally authorized substrates as feed for insects (i.e. Reg. 1069/2009 & Reg. 142/2011) and Regulation (EC) No 767/2009 (on the placing on the market and use of feed), which forbids the use of manure/animal feces. Table 3 reviews substrates used within the EU. Substrate suppliers need to be audited to guarantee the safety of their products. Procedures related to safety management, production, hygienic practices, storage conditions and implementation of HACPP procedures need to be registered to allow traceability. Specifications for composition and conditions of usage need to be recorded (IPIFF 2019).

Seaweed meal is recognized as an animal feed substrate in the EU (Reg (EC) 68/2013; EC 2013) and for the production of insect larvae (Liland et al.

Table 3 List of main used and non-authorized substrates for insect feeding

Used	Non-authorized
Cereal-based materials	Human excrements
Fruits and vegetables and their derived products	Livestock manure and/or slurry or other products derived from animal digestive tract content
Commercial feed	Water treatment (industrial sludges) or solid waste streams (urban industrial or domestic)
Unsold products (due to 'technical defects') from supermarkets, food industry or bakery establishments (the so-called 'former foodstuffs' in accordance with EU legislation) if containing materials of vegetal origin, eggs and/or milk products	Animal by-products originating from slaughterhouses or rendering establishments (e.g. processed animal proteins)
	Seeds treated with plant protection products
	Treated wood
	Food waste from restaurants, catering establishments and household
	'Former foodstuffs' containing materials of animal origin (excluding eggs, milk and their derived products)
	Packaging and parts thereof (plastic, PET, paper)

Source: IPIFF 2019.

2017). However, there are no microbiological standards for seaweed meals in the EU, due to a limited number of studies in this field. Swinscoe et al. (2020) demonstrated that risk assessment must consider not only the bacterial hazards associated with these novel ingredients, but also the behavior of those bacteria submitted to abiotic and biotic processing stresses. This knowledge will help in the improvement of HACCP guidelines, good agricultural practices (GAP) and GMP.

4 Managing feed risks: assessment, management, and communication

Feed risk analysis consists of three stages:

i risk assessment;
ii risk management; and
iii risk communication.

Risk assessment of feed contaminants has a critical role in determining reference thresholds for prevention and intervention. When a contaminant is identified as a potential risk, key requirements include data on the composition of the hazard under analysis, available methods to analyze the hazardous compound in feed and appropriate toxicological data. Based on this, it is possible to set safe limits for the compound in animal feed. It is important to note that substances found in feed have two target populations: animals that are directly exposed and consumers of edible animal products (Mantovani and Frazzoli 2010).

During feed production, different mixtures pass through the same production line. The remainder of the first batch of feed produced is mixed with the subsequent feed batch produced. This process cannot be avoided and carry-over of contaminants may occur (Radulovic et al. 2019). This is a particular concern in countries where high use of veterinary drugs via animal nutrition is allowed, resulting in cross contamination of feed with several active ingredients (Barbosa da Silva et al. 2019). The main sources of carry-over occurring in the feed mill are presented in Table 4.

Risk management is based on the results of a risk assessment when the limits for levels of contaminants are set. The government is responsible for monitoring feed supply and the use of the approved compounds to ensure they are within legal limits, while feed producers are responsible for applying and administering risk management quality and safety systems (Lupien 2007). This chapter focuses on risk management for the prevention and control of feed contaminants (see Section 5).

The final stage, risk communication, consists of the technical information shared between government, feed producers, academia, legislators and the

Table 4 Sources of carry-over regarding feed plant equipment

Equipment	Mode of carry-over
Dust system	Delayed return of dust to production line
	Excessive pickup of drug and carrier
	Han-up (electrostatic or moisture)
Mixer	Residual feed remaining in mixer
	Buildup of material on ribbons and walls
	Electrostatic hang-up on walls and top
	Leaking miker gate (not fully closed)
Surge bin	Incomplete clean-out
	Han-up (electrostatic or moisture)
Conveyors	Same as surge bin
Elevators	Residual feed remaining in buckets and boot
	Han-up (electrostatic or moisture)
Bins	Bridging
	Residual feed from incomplete clean out
Bulk truck	Error in bin chart records
	Incomplete clean out
	Bridging and hang-up

Source: Radulovic et al. 2019.

public. The main goal of this communication is to agree on an overall strategy and policy for managing risk. To achieve this goal requires engaging different sectors in society to create public understanding of choices in managing animal feed contamination risks. Risk Communication Advisory Committees are necessary to achieve this objective. In the United States the FDA Food Advisory Committee evaluates and designs strategies and programs about the risks and benefits of FDA-regulated products to communicate with the public. The EFSA Scientific Panel not only gathers information from other experts, but also collects information from the feed industry and conducts colloquia to exchange information.

5 Risk management for feed contaminants: measures and challenges

To enter the market, feed must be safe for animal consumption and must not have any adverse effects on human health. To guarantee this, all stages of feed production chain must follow safety protocols which need to fulfill different aspects of risk management. Risk analysis plays a key role at the beginning of the process in identifying food safety problems so that a risk management system can subsequently be developed. Key conditions to perform a risk analysis are:

- Operational feed safety system: a national feed control strategy is needed to provide the scientific and technical foundation to implement laws and regulations;
- Knowledge for risk analysis: this step is the responsibility of government officials and decision-makers; and
- Support and participation of key stakeholders: a risk analysis will be effective only if government, industry, academic institutions and consumers recognize the importance of participating in the process.

Meeting these conditions provides the framework for effective risk management.

Risk management has been defined as 'the process, distinct from risk assessment, of weighing policy alternatives, in consultation with all interested parties, considering risk assessment and other factors relevant for the health protection of consumers and for the promotion of fair trade practices, and, if needed, selecting appropriate prevention and control options' (Codex). There are many challenges in designing risk management measures for preventing undesirable substance in animal feed, as highlighted by Verstraete (2013). Challenges include:

- lack of data in such areas as toxicity, carry-over of contaminants from feed to the food of animal origin, and accurate data on occurrence in feed and food;
- lack of availability of reliable routine methods of analysis;
- varying sensitivity of animals to a particular undesirable substance, depending on animal species, age and gender;
- the problem that carry-over from feed to the food of animal origin is animal species-specific and might be different for different foods of animal origin (muscle tissue, edible offal, eggs and milk);
- the fact that feed materials are not marketed for specific species; this makes it very difficult to develop risk management measures that can take into account animal species-specific sensitivity to undesirable substances in feed materials;
- the need to market and label compound feed for specific animal species so that risk management measures for compound feed can take into account animal species sensitivity to particular undesirable substances;
- the fact that bio-availability of an undesirable substance present in a feed may vary according to the nature of the feed. For example, metals are less

bio-available when present in feed of mineral origin than when present in feed of plant origin;

- the issue that feed materials include a very wide range of products, not just by-products of food production but also by-products of other production processes such as the production of biofuel;
- the fact that it is not always straightforward to determine what level can be achieved under normal good practice production conditions worldwide; and
- the problem that, in some cases, feed use is the major use of a certain feed material; in cases of non-compliance, this means there are limited alternative uses for such feed material. Available options for detoxification also have to be considered.

This list illustrates the range of challenges in achieving effective risk management. It is clear therefore that risk management is not a simple linear process but much more complex, requiring flexible models depending on the hazard and animal species. Each country usually has its own management models and procedures.

6 Risk management systems

To reduce feed contamination at different stages of food production (e.g., feed production, on the farm, during slaughter and further processing of food products), it is necessary to implement a quality management process to manage risks. According to Ma et al. (2005), the process of risk management takes place both before the occurrence of a contamination problem and after it has occurred. The first objective is to reduce potential risks, while the second objective is to correct the problem. To meet these aims, prevention and control systems must be developed and constantly evaluated. The risk management of feed contaminants relies primarily on the implementation of good practice in feed production. Based on a risk assessment, different systems can be applied to focus on the various possible entry points for contaminants. In practice there is a variety of systems, some of which are based on official guidance and procedures whilst other systems are simultaneously applied by the feed business industry developing its own internal systems for quality control.

6.1 Current Good Manufacturing Practices (cGMP)

In 1965 the FDA implemented the current Good Manufacturing Practices (cGMP) regulations. The cGMP requirements serve as guidelines for medicated feed manufacturers to ensure that the final product meets quality standards when entering the food chain. The FDA applies two categories to medicated feed:

- Category I for drugs that do not require a withdrawal time; and
- Category II which encompasses drugs that either require a withdrawal period or which have an established zero tolerance for the residual limit in the edible tissue.

These categories are further divided into the following types:

- Type A: a medicated substance that contains the most concentrated form of the drug. It cannot be used for feed but can be mixed with feed ingredients to produce Type B or C medicated feed;
- Type B: a medicated animal feed containing a new animal drug that is intended solely for manufacture into another Type B- or Type C-medicated feed, and is produced from a drug component, a Type A-medicated article, or another Type B-medicated feed; and
- Type C: before being fed to animals, the Type B-medicated feed has to be substantially diluted with one or more nutrients to produce Type C-medicated feed, that is, a medicated animal feed containing a new animal drug that may be offered as a complete feed.

It is important to note that Type A-medicated articles and Type B- and Type C-medicated feeds have differential requirements for approval, manufacture and registration. There is also a difference between medicated and animal drugs. Medicated feed is a feed containing one or more substances considered as an animal drug by the FDA, while an animal drug includes substances used for prevention or treatment of diseases.

If a medicated feed is produced in a facility that does not meet applicable cGMP standards, the FDA will consider medicated feed as adulterated, even if the feed mill is a FDA-licensed factory. The cGMP regulations do not contain specific instructions but describe the minimum requirements that must be met by the feed manufacturer (Scheid 2012). The cGMP regulations can, however, be summarized as followed for feed mills (Boyd 1994):

1 Good housekeeping inside the facility and throughout the surrounding area;
2 Buildings and equipment that are suitable in all respects for the production of medicated feeds;
3 Personnel that have adequate training and experience with appropriate supervision;
4 Operation of facilities in a planned and systematic manner from receipt of animal drugs to shipment of medicated feeds;

5 Controls that provide for product integrity and prevent unsafe carry-over of drugs into other production lines; and
6 Records that provide a meaningful written history of the entire production process and permit tracing of product to and from the field.

These points can be categorized into four important areas:

1 Inventory control: with planned and actual physical inventories of products that need to match with each other. Any significant discrepancy between them should result in a hold on the distribution of the product in question and an immediate investigation to determine whether or not a problem exists. The cause of the discrepancy should be determined and corrective action taken.
2 Records: The system of records, starting with drug receipt through production to distribution, should provide an accurate history of production for both current and future uses.
3 Carry-over control: A basic premise of the cGMP regulations is maintaining the integrity of all feeds, medicated and non-medicated, produced in the same facility. However, one cannot exclude the risk of carry-over of a contaminant from one feed to another. cGMP therefore requires the prevention of unsafe carry-over of contaminants. This can be achieved, for example, by grouping production of feeds containing the same drug together. Production must also start at the highest concentration and finish with the lowest.
4 Labeling: All feeds intended for distribution must legally be labeled with specific information concerning their contents, including drug name, amount, purpose or claim, directions for use, and all required precautions or warnings.

6.2 The US Animal Feed Safety System (AFSS)

In 2003 the FDA implemented the US Animal Feed Safety System (AFSS), followed by the Food Safety Modernization Act (FSMA) in 2011, which changed the regulatory emphasis to prevention rather than simply reacting to feed safety problems. The FDA manages the safety of animal feed under the AFSS together with the Center for Veterinary Medicine (CVM). The AFSS, in its turn, covers the aspects of animal feed safety shown in Table 5.

All animal feed regulation is under AFSS supervision. This covers animal feed manufacturers (facilities and equipment involved in production, packaging, storage and distribution of animal feed and animal feed ingredients). The FDA lists all approved additives for animal feed in Document 21 CFR 573.

Table 5 Aspects of animal feed safety according to AFSS

Aspects	Description
Regulatory activities	Pertaining to all aspects of animal feed safety – pre-approval requirements, establishing limits on feed hazards, manufacturing, labelling, packaging, storing, and distributing
Enforcement	Performing inspections and investigations; Taking enforcement actions to remove unsafe animal feed from the marketplace and to address firms failing to comply with FDA regulations
Information dissemination	Providing education and training, conducting research
Cooperation	Establishing partnerships with other government agencies with responsibility for feed safety AFSS

Source: FDA.

Although animal feed ingredient suppliers can use the Generally Recognized as Safe (GRAS) Notification Procedure to use ingredients that fall within GRAS parameters, the FDA does not give GRAS-related approval for intended conditions of use.

For production, storage and distribution of safe animal feed and ingredients, animal feed producers must comply with cGMP and the hazard analysis and risk-based preventive controls are required by the FSMA. The FDA has developed a 'Guidance for Industry' document addressing hazard analysis and preventive measures. Knowing that animal feed hazards carry variable risk levels, a ranking system is used which is regularly updated, taking into account animal species, animal category and type of diet. New legislation focused on prevention has provided by new tools focused on the recall of products, access to records, and the ability to track and trace domestic and imported animal feed. Specific responses to hazards are given within the complete feed process.

6.3 The hazard analysis critical control point (HACCP) system

The hazard analysis critical control point (HACCP) system is a methodology that identifies, evaluates and control hazards (physical, chemical, biological and allergenic), focusing on a proactive approach. HACCP principles are designed to carry out the following:

a Identify any hazards that must be prevented, eliminated or reduced to acceptable levels;
b Identify the critical control points at the step or steps at which control is essential to prevent or eliminate a hazard or reduce it to acceptable levels;

c Establish critical limits at critical control points which separate the acceptable from unacceptable, for the prevention, elimination or reduction of identified hazards;
d Establish and implement effective monitoring procedures at critical control points;
e Establish corrective action when monitoring indicates that a critical control point is not under control;
f Establish procedures to verify that the measures outlined in points (a) to (e) are complete and effective. Verification procedures must be carried out regularly; and
g Establish documents and records commensurate with the nature and size of the feed business to demonstrate the effective application of the measures set out in (a) to (f).

When implementing a HACCP management system, management must follow steps such as:

i defining the scope of the feed safety management system;
ii identifying products and production sites covered by the system and ensuring the establishment of safety objectives;
iii guaranteeing that feed safety is a core aim of the business; and
iv reviewing the feed safety system at least once a year. This review should also occur when feed mill plant is changed.

Feed business operators usually produce different diets for a range of animal species within the same establishment and production line. This brings the risk that traces of a compound remain in the production line and enter production of subsequent or non-target feed, resulting in carry-over or cross-contamination (Verstraete 2013).

It is important to note that the HACCP system does not replace official controls and is focused on the product, separating acceptable from unacceptable product. One of the examples of this system is risk management of carry-over in the feed mill. There are many feed safety standards in different countries providing guidance on how to apply HACCP in the feed sector (Table 6) as reviewed by Bouxin (2014). HACCP systems take into account carry-over risks, characteristics of hazards, and the animal species fed with the authorized compounds. The HACCP system can also be optimized by advances in process analysis technology (PAT), which can be applied in each step of HACCP management systems and will result in a direct, continuous and rapid measurement of biological, chemical and physical risks during feed production and processing (Moran et al. 2017).

Table 6 Feed safety management documents, application, country of origin and area of application, and type of tool providing HACCP guidance

Name	Application					Country	Area	Tool
	FM	FA	PR	CF	TR			
SFSF	x	x	x	x		USA	USA	CS
EFMC			x	x		EU	EU	Code
GMP+ International	x	x	x	x	x	NL	Global	CS
QS	x	x	x	x	x	DE	DE	CS
FEMAS	x	x				UK	Global	CS
UFAS			x	x		UK	UK	CS
Oqualim			x	x		FR	FR	CS
Ovocom	x	x	x	x	x	BE	BE	CS
EFISC	x					EU	EU	CS
GTP					x	EU	Global	CS
FAMI-QS		x				EU	Global	CS
Feed & Food Safety	x	x	x	x		BR	BR	CS
GLOBAL GAP				x	x	Global	Global	CS

Source: Bouxin 2014.
EFISC: European Feeds Ingredients Safety Certificering; EFMC: European *Feed* Manufacturers' Guide; FA: Feed additives; FAMI-QS: Feed Additive and preMIxture Quality System; FEMAS: *Feed* Materials Assurance Scheme; FM: Feed material; GLOBAL GAP: Global Good Agricultural Practices; GMP: Good Manufacturing Practice; GTP: Good Trading Practices; PR: Premixes; QS: Quality Scheme for safe food; SFSF: Safe Feed/Safe Food; TR: Trade; UFAS: Universal *Feed* Assurance Scheme.

6.4 GMP+ Feed Safety Assurance

By adapting the HACCP system, the Dutch animal feed sector has implemented a quality assurance system that is also applied in the broader European feed industry. The system underpins the feed certification schemes for Feed Safety Assurance (GMP + FSA) and Feed Responsibility Assurance (GMP + FRA). The GMP+ FSA certificate is focused on risk management, while GMP+ FRA certifies that companies meet market requirements for human and environmental safety. There are currently more than 18 000 participating companies in more than 70 countries, explaining why GMP+ International is a leading global player in feed safety.

The GMP+ FSA scheme consists of four categories of documents:

i general governance documents;
ii normative standards;
iii rules of certification, supervision and monitoring; and
iv informative documents like guides and reports.

The main reasons why many companies implement this certificate system are:

- Its focus on quality (ISO 9001) and risk (HACCP) management;
- access to supporting information and services, such as risk assessments, fact sheets, monitoring data, early warning messages, and so on;
- guarantees of the continuity of business operations, because the corporate structure is set up more efficiently and costs and financial risks are limited;
- improvements to company image and reliability;
- making certified companies part of a globally operating feed safety system;
- acceptance by other international schemes;
- the possibility of combining multiple certifications (e.g. combining feed safety and sustainability); and
- the contribution to uniformity in the feed chain.

Due to its pro-active and preventive rather than reactive nature, a GMP+ system is a chain-focused approach. This means that feed safety is controlled from the field onwards to avoid further contamination of animal diets. When compared to EU guidelines, this system is usually stricter on mycotoxins levels than limits recommended by governments in the final diet of livestock. An exception is observed for levels of the mycotoxin Ochratoxin A (OTA) in poultry diets. Instead of indicating a single limit per mycotoxin, GMP+ considers two limits: action and rejection limits. The limits for mycotoxins in complete feed were revised in 2019 as shown in Table 7.

The GMP+ FSA scheme is reviewed by Hartog (2012). In brief, the most important characteristics are:

1. Quality management following ISO 9001: Quality management here involves product quality and the methods applied to reach this quality, thus ensuring effective process control;
2. HACCP and risk assessment: Since 2000 the, HACCP system has been integrated into the GMP+ certification scheme, which makes the scheme a preventive (rather than reactive) approach to feed safety management; and
3. Product standards: these refer to the maximum permitted levels of contaminants such as chemical and microbiological agents. Usually, these levels follow EU standards, but some of them are based on requirements in the feed chain. In the case of mycotoxins, for instance, guidance levels in GMP+ FSA are stricter than those found in EU guidelines (Table 7).

Table 7 List of maximum guidance levels of mycotoxins in complete diet as recommended by EU, US and GMP+

				GMP+ (ppm)	
Mycotoxin	Species	EU (ppm)	US (ppm)	Action limit	Rejection limit
Aflatoxin	Piglets	0.005	0.020[1]	-	0.005
	Cattle	0.020	0.030[1]	-	0.020
	Calves < 4mo	0.005	0.020[1]	-	0.005
	Dairy cattle	0.005	0.020[1]	-	0.005
	Young poultry	0.005	0.020[1]	-	0.005
	Poultry	0.020	0.020	-	0.020
DON	Pigs	0.9	1.0	0.8	1.0
	Cattle	5.0	10.0	4.0	5.0
	Calves < 4mo	2.0	-	1.6	2.0
	Dairy cattle	5.0	5.0	2.4	3.0
	Poultry	5.0	5.0	3.2	4.0
ZEA	Piglets e gilts	0.10	-	0.08	0.10
	Sows	0.25	-	0.20	0.25
	Fattening pigs	0.25	-	0.20	0.25
	Cattle	0.50	-	0.40	0.50
OTA	Pigs	0.05	-	0.04	0.05
	Poultry	0.10	-	0.16	0.20
FB1 + FB2	Pigs	5.0	20.0	5.0	
	Cattle	50.0	30.0	50.0	
	Calves < 4mo	20.0	-	20.0	
	Beef cattle	50.0	60.0	50.0	
	Poultry	20.0	100[2,3]	20.0	
Ergot	-	-	-	-	1.0

European maximum levels (Recommendation 2006/576/EC); US maximum levels (FDA); [1]Corn-based diets; [2]Maximum level for broilers; [3]breeding poultry (30 ppm).

4 Prerequisite programs (PRPs) for animal feed manufacturing: Although risk assessment is related to HACCP, the focus is also put on all PRPs and technologies that work together with the HACCP system. PRPs include good hygiene practices (GHPs), standard operating procedures (SOPs) and sanitation standard operating procedures (SSOPs).

5 Chain approach: This approach follows the basic principle that every company in the feed chain takes responsibility for the safety of its

supplied services and products in ensuring the safety of the final product.

6 Monitoring: Regular sampling and testing are crucial to verify the effectiveness of control measures, as well as to detect contaminants in an early stage of feed production. GMP+ FSA indicates specific sampling methods and quality requirements for laboratories.
7 Traceability: All incoming and outgoing goods must be registered to allow tracking and tracing of possible contaminated feed materials, avoiding the spread of a contaminant and allowing the recall of goods when necessary.
8 Early warning system (EWS): A GMP+ certified company detecting contamination that may affect the other participants is obliged to inform GMP+ International. For example, when a feedstuff is analyzed and its mycotoxin levels are higher than the allowed limit, the EWS is used to inform all participants which feedstuff is contaminated (batch, mycotoxins and levels) together with the information on the levels allowed.

In 2018, GMP+ International started renewing its scheme standards whilst maintaining the same basic content. The changes are aimed to be concluded in 2020 and will include a more simplified scheme, *Core standard.* Documents will also contain explanations and examples to support participants (www.gmpplus.org).

7 Summary and future trends

Feed contaminants are often unavoidable compounds due to the different sources of contamination, for example, from the field, additives, accumulated environmental pollutants, storage contaminants, and those as a result of cross-contamination, which then pose a food safety risk. There is still a significant lack of toxicity and occurrence data to enable the design of a broad risk assessment of the presence of a number of undesirable substances in feed. With advances in analytical techniques, this gap has increased since more compounds are being detected. Risk assessment and management remain a challenge for mycotoxins in particular because most feedstuffs are contaminated with multiple mycotoxins and the interaction of these fungal toxins in the animal diet needs to be considered.

Systems to analyze and control risks for feed contaminants vary from country to country, which makes harmonization of legislation and the implementation of global good practices in feed production challenging. However, the application of systems that ensure the quality of ingredients, appropriate transport, and feed production, as well as the correct use of additives is crucial to guarantee a safe feed and edible animal product.

When discussing management systems, it is important to point out the crucial importance of prevention. Prevention will be always less costly than control and mitigation after the hazard enters the feed chain. This step should be applied to identify risks at the earliest stage as possible, avoid contamination in the first place and decrease carry-over during feed processing. This requires knowledge of the feed chain and feed mills' processing systems as well as collaboration among operators.

Global trading requires harmonized risk analysis systems and efficient management systems. With advances in analytical techniques, online training, and virtual visits to the plant where feed is processed, systems will be optimized and risks of feed contamination can be decreased. Medicated feed is becoming better controlled in response to the requirements of both scientific research and the requirements of consumers. Preventative management systems are focused on avoiding undesirable drugs in diets and, when they are allowed, carry-over contamination can be significantly decreased by intelligent protocols.

8 Where to look for further information

Regulations and guidance's from Europe and the United States can be found below.

- Code of Federal Regulations FDA. US Food and Drug Administration. Available online: https://www.accessdata.fda.gov/scripts/cdrh/cfdocs/cfcfr/CFRSearch.cfm?CFRPart=573&showFR=1&subpartNode=21:6.0.1.1.22.2.
- Compliance Policy Guide (CPG) Sec. 680.600 – Sequencing as a means to prevent unsafe drug contamination in the production, storage, and distribution of feeds. Available online at: https://www.fda.gov/regulatory-information/search-fda-guidance-documents/cpg-sec-680600-sequencing-means-prevent-unsafe-drug-contamination-production-storage-and.
- Directive 2002/32/EC addresses a range of undesirable substances, such as the classical pesticides, fluorine, heavy metals, gossypol and dioxins. Another regulation applies to residues of plant protection products (Regulation (EC) 396/2005).
- European Commission (EC) (2013). Commission Regulation (EU) 68/2013/EC of 16 January 2013 on 652 the Catalogue of feed materials. *Official Journal of the European Union* L29: 1–64.
- Food and Agriculture Organization (FAO) (2015a). Slaughterhouses. FAO Corporate Document Repository. Prepared by Agriculture and Consumer Protection. Food and Agriculture Organization.
- Food and Agriculture Organization (FAO) (2015b). Animal Production and health. Source of meat. Available online at: http://www.fao.org/ag/againfo/themes/en/meat/backgr_sources.html.

- International Platform of Insect for Food and Feed. Guide on Good Hygiene Practices for European Union (EU) Producers of Insects as Feed and Food. Available online at: http://ipiff.org/wp-content/uploads/2019/03/IPIFF_Guide_A4_2019-v5-separate.pdf (accessed on 13 May 2020).
- GMP+ international BA2 Feed Certification scheme, Control of residues, Version EN:1 July 2018. Available online at: https://www.gmpplus.org/media/2276/gmpplus-ba2-en-20180701.pdf.
- Regulation EU 2019/4 of the European Parliament and of the Council of 11 December 2018 on the manufacture, placing on the market and use of medicated feed, amending Regulation (EC) No 183/2005 of the European Parliament and of the Council and repealing Council Directive 90/167/EEC, Official Journal of the European Union.
- Regulation (EC) No 183/2005 of the European Parliament and of the Council of 12 January 2005 laying down requirements for feed hygiene. Off. J. Eur. Union, L 35, 8.2.2005, p. 1. Available online at: http://eur-lex.europa.eu/LexUriServ/LexUriServ.do?uri=OJ:L:2005:035:0001:0022:EN:PDF.

For more information on animal feed contamination, a book is recommended:

- Fink-Gremmels, J. (2012). Animal feed contamination: effects on livestock and food safety. Woodhead Publishing Limited.

9 References

Alm, M. (2012). Review of the EU feed ban on non-ruminant processed animal proteins: outlook of the European Fat Processors and Renderers Association (EFPRA). TAIEX Workshop, 27th November 2012, Stavropol (Russia).

Barbosa da Silva, A., Back, M., Daguer, H., Palmeira, M., De Sá Ploêncio, L. A., Molognoni, L., Peripoli, V. and Bianchi, I. (2019). Carry-over and contamination of veterinary drugs in feed production lines for poultry and pigs. *Food Addit. Contam. A* 36(5): 740–51.

Battilani, P., Palumbo, R., Giorni, P., Dall'Asta, C., Dellafiora, L., Gkrillas, A., Toscano, P., Crisci, A., Brera, C., De Santis, B., Cammarano, R., Seta, M. D., Campbell, K., Elliot, C., Venancio, A., Gonçalves, A., Terciolo, C. and Oswald, I. (2020). Mycotoxin mixtures in food and feed: holistic, innovative, flexible risk assessment modelling approach: MYCHIF. EFSA Supporting Publication 2020: EN-1757. 161 pp.

Bosch, G., Fels-Klerx, H. J. V., Rijk, T. C. and Oonincx, D. G. A. B. (2017). Aflatoxin b1 tolerance and accumulation in black soldier fly larvae (*Hermetia illucens*) and yellow mealworms (*Tenebrio molitor*). *Toxins* 9(6): 185.

Bouxin, A. (2014). Management of safety in the feed chain. In: Motarjemi, Y., Lelieveld, H. and PROSE Awards (Eds), *Food Safety Management–A Practical Guide for the Food Industry*. Academic Press, 23–43.

Boyd, L. H. (1994). Regulation of drugs and chemicals used by the poultry industry. Good manufacturing practices. *Poult. Sci.* 73(9): 1419–22.

Charlton, A. J., Dickinson, M., Wakefield, M. E., Fitches, E., Kenis, M., Han, R., Zhu, F., Kone, N., Grant, M., Devic, E., Bruggeman, G., Prior, R. and Smith, R. (2015). Exploring the chemical safety of fly larvae as a source of protein for animal feed. *J. Insects Feed Food* 1(1): 7–16.

Cheli, F., Pinotti, L., Rossi, L. and Dell'Orto, V. (2013). Effect of milling procedures on mycotoxin distribution in wheat fractions: a review. *Food Sci. Technol.* 54(2): 307–14.

den Hartog, J. (2012). The GMP+ feed safety assurance (FSA) scheme. In: Fink-Gremmels, J. (Eds), *Animal Feed Contamination: Effects on Livestock and Food Safety*. Oxford: Elsevier, 625–49.

FAO. (2020). Global Feed Safety Platform. Available at: http://www.fao.org/feed-safety/background/why-feed-safety/en/.

German Federal Institute for Risk Assessment (BfR), National Reference Laboratory for Animal protein in Feed, NRL-AP (BfR), National Reference Laboratory for Animal protein in Feed, NRL-AP, Garino, C., Zagon, J. and Braeuning, A. (2019). Insects in food and feed–allergenicity risk assessment and analytical detection. *EFSA J.* 17(Suppl 2): e170907.

Gerssen, A., Bovee, T. H. F., van Ginkel, L. A., van Iersel, M. L. P. S. and Hoogenboom, R. L. A. P. (2019). Food and feed safety: cases and approaches to identify the responsible toxins and toxicants. *Food Control* 98: 9–18.

Gordon, R. K., Kotowski, I. K., Coulson, K. F., Link, D., MacKenzie, A. and Bowling-Heyward, J. (2019). The role of non-animal origin feed ingredients in transmission of viral pathogens of swine: a review of scientific literature. *Front. Vet. Sci.* 6: 273.

Hejna, M., Gottardo, D., Baldi, A., Dell'Orto, V., Cheli, F., Zaninelli, M. and Rossi, L. (2018). Review: nutritional ecology of heavy metals. *Animal* 12(10): 2156–70.

International Platform of Insect for Food and Feed. Guide on Good Hygiene Practices for European Union (EU) Producers of Insects as Feed and Food. Available at: http://ipiff.org/wp-content/uploads/2019/03/IPIFF_Guide_A4_2019-v5-separate.pdf (Accessed on 13 May 2019).

Jaw, Y. M., Chen, Y. Y., Lee, Y. C., Lee, P. H., Jiang, C. M. and Tsai, Y. H. (2012). Histamine content and isolation of histamine-forming bacteria in fish meal and fish soluble concentrate. *Fish. Sci.* 78(1): 155–62.

Jedrerek, D., Levic, J., Wallace, J. and Oleszek, W. (2016). Animal by-products for feed: characteristics, European regulatory framework, and potential impacts on human and animal health and the environment. *J. Anim. Feed Sci.* 25(3): 189–202.

Lähteenmäki-Uutela, A., Grmelova, N., Henault-Ethier, L., Deschamps, M. H., Vandenberg, G. W., Zhao, A., Zhang, Y., Yang, B. and Nemane, V. (2017). Insects as food and feed: laws of the European Union, United States, Canada, Mexico, Australia, and China. *Eur. Food Feed Law Rev.* 12: 22–36.

Leeman, W. R., Van den Berg, K. J. and Houben, G. F. (2007). Transfer of chemicals from feed to animal products: the use of transfer factors in risk assessment. *Food Addit. Contam.* 24(1): 1–13.

Liland, N. S., Biancarosa, I., Araujo, P., Biemans, D., Bruckner, C. G., Waagbo, R., Torstensen, B. E. and Lock, E. J. (2017). Modulation of nutrient composition of black soldier fly (*Hermetia illucens*) larvae by feeding seaweed-enriched media. *PLOS ONE* 12(8): e0183188.

Lupien, J. R. (2007). Prevention and control of food safety risks: the role of governments, food producers, marketers, and academia. *Asia Pac. J. Clin. Nutr.* 16 (Suppl. 1): 74–9.

Ma, S. H., Lin, Y. and Chen, Z. X. (2005). *Supply Chain Management*. Beijing: Higher Education Press.

Mantovani, A. and Frazzoli, C. (2010). Risk assessment of toxic contaminants in animal feed. *CAB Rev.* 5(46): 1–14.

Moran, F., Sullivan, C., Keener, K. and Cullen, P. (2017). Facilitating smart HACCP strategies with process analytical technology. *Curr. Opin. Food Sci.* 17: 94–9.

Morrow, J. D., Margolies, G. R., Rowland, J. and Roberts, L. J. (1991). Evidence that histamine is the causative toxin of scombroid fish poisoning. *N. Engl. J. Med.* 324(11): 716–20.

Pandey, P., Vidyarthi, S. K., Vaddella, V., Venkitasamy, C., Pitesky, M., Weimer, B. and Pires, A. F. A. (2020). Improving biosecurity procedures to minimize the risk of spreading pathogenic infections agents after carcass recycling. *Front. Microbiol.* 11: 623.

Radulovic, S., Sefer, D., Markovic, R., Lukic, M., Nikolic, A. and Milicevic, D. (2019). Risk assessment and risk management of contaminants in the feed to food chain. *IOP Conf. Ser.: Earth Environ. Sci.* 333: 012030.

Ruiz-Capillas, C. and Herrero, A. M. (2019). Impact of biogenic amines on food quality and safety. *Foods* 8(2): 62.

Scheid, J. F. (2012). Management of animal feed safety in the USA. In: Fink-Gremmels, J. (Eds), *Animal Feed Contamination: Effects on Livestock and Food Safety*. Oxford: Elsevier, 608–23.

Schoevers, E. J., Santos, R. R., Colenbrander, B., Fink-Gremmels, J. and Roelen, B. A. J. (2012). Transgenerational toxicity of zearalenone in pigs. *Reprod. Toxicol.* 34(1): 110–9.

Schrögel, P. and Wätjen, W. (2019). Insects for food and feed-safety aspects related to mycotoxins and metals. *Foods* 8(8): e288.

Silva, T. E. S., de Brito, D. C. C., de Sá, N. A. R., da Silva, R. F., Ferreira, A. C. A., da Silva, J. Y. G., Guedes, M. I. F., Rodrigues, A. P. R., Santos, R. R. and de Figueiredo, J. R. (2019). Equol: a microbiota metabolite able to alleviate the negative effects of zearalenone during in vitro culture of ovine preantral follicles. *Toxins* 11(11): 652.

Sogari, G., Amato, M., Biasato, I., Chiesa, S. and Gasco, L. (2019). The potential role of insects as feed: a muulti-perspective review. *Animals* 9(4): e119.

Swinscoe, I., Oliver, D. M., Ornsrud, R. and Quiliam, R. S. (2020). The microbial safety of seaweek as a feed component for black soldier fly (*Hermetia illucens*) larvae. *Food Microbiol.* 91: 103535.

van Raamsdonk, L. W. D., Prins, T. W., Meijer, N., Scholtens, I. M. J., Bremer, M. G. E. G. and de Jong, J. (2017). Bridging legal requirements and analytical methods: a review of monitoring opportunities of animal proteins in feed. *Food Addit. Contam. A* 36(1): 46-73.

Verbeek, W., Spranghers, T., De Clercq, P., De Smet, S., Sas, B. and Eeckhout, M. (2015). Insects in animal feed: acceptance and its determinants among farmers, agriculture sector stakehoulders and citizens. *Anim. Feed Sci. Technol.* 204: 72–87.

Verstraete, F. (2013). Risk management of undesirable substances in feed following updated risk assessments. *Toxicol. Appl. Pharmacol.* 270(3): 230–47.

Wales, A. D., Allen, V. M. and Davies, R. H. (2010). Chemical treatment of animal feed and water for the control of *Salmonella*. *Foodborne Pathog. Dis.* 7(1): 3–15. Available online at: www.gmpplus.org.

Chapter 8

Developing effective product dossiers for regulatory approval of new animal feed products

Manfred Lützow, saqual GmbH, Switzerland

1 Introduction: what are feed products?

Feed is part of the food supply chain on which society relies for production not just of food for humans but also for pets and non-food animal products such as wool from sheep. The scale of feed production (as part of the overall food supply chain) can be seen in the 164.811 million tons of compound feed used in the 28 member states of the European Union (EU) in 2018 (FEFAC, 2019). In 2018, 321.3 kg of compound feed was produced for each citizen of the EU.

Countries differ in their food supply systems and therefore have different regulatory frameworks to ensure that sufficient safe food of the required quality and quantity is available. Regulatory differences potentially hinder trade and, as a result, there have been significant efforts to harmonise food legislation. The Codex Alimentarius Commission, an international non-governmental

http://dx.doi.org/10.19103/AS.2020.0083.15

organisation that is managed jointly by the Food and Agriculture Organisation of the United Nations (FAO) and the World Health Organisation (WHO), has been developing food standards for almost 60 years to address key aspects of food quality and safety. Many countries have implemented Codex standards, their food legislation and regulatory frameworks are similar, they use the same definitions and legal terms and have comparable approaches. We have for example internationally accepted definitions for food additives or the requirements for the composition and labelling of dairy fat spreads.

Such international regulatory harmonisation is less advanced in the area of feed. The Codex Alimentarius Commission started the discussion of animal feed much later than for food and, so far, has not developed any binding overall rules governing the international trade in feed. A Codex task force has developed general principles for the risk assessment of animal feed (Codex, 2013). However, no international feed product-specific standards have been developed (apart from the adoption of common standards for certain contaminants in feed e.g. for some mycotoxins). In the absence of internationally accepted definitions, the feed product categories and terms applied by the relevant jurisdiction need to be used. Those are, unfortunately, quite heterogeneous (Smedley, 2013). In part they reflect differences in responsibility for feed safety. In the United States for example the Food and Drug Administration (FDA) regulates food additives. This means there are no feed additives as such but, rather food additives intended for animal feed. There are also differences in whether a product constitutes feed. Animals do not only receive feed but are also treated with various other products such as veterinary products and biocides which may or may not be defined as feed. The main categories defined by Codex Alimentarius, the United States and the European Union are compared in Table 1.

Prior to the development of a dossier for a feed product, the regulatory framework of the target market needs to be understood, as requirements differ significantly between jurisdictions. The following outline is based on the regulatory framework of the European Union that has been developed over the past 20 years and revised recently from 2017 to 2019. No other jurisdiction has gone so far in incorporating current risk assessment principles and

Table 1 Comparison of categories and terms for feed products (Codex, 2013; Smedley, 2013)

Codex Alimentarius	United States	European Union
Feed (Feeding stuff)	Animal Feed	Feed
Feed additive	Food Additive	Feed additive
Feed ingredient	Feed Ingredients	Feed material
Veterinary drug	Animal Drugs (Medicated feeds)	Veterinary medicinal product
-	Utility, claim	Effect, function, claim

state-of-the-art scientific methodology in their official guidance documents. This does not mean that other agencies involved in the evaluation of feed ingredient dossiers are not using state-of-the-art methodology, but their published guidance is often less up-to-date. It should be noted that the life-cycle approach to the evaluation of feed ingredients discussed in the following section is used by most jurisdictions.

2 The food chain: the product life cycle for feed ingredients

Animal feed is composed of agricultural raw materials such as wheat middlings, rice bran, soybean meal, minerals such as calcium phosphate, trace minerals like manganese salts, vitamins such as retinyl acetate (vitamin A), sources of proteins and amino acids, carbohydrates and lipids (oils and fats). Feed also includes ingredients that serve a specific purpose such as preservatives (benzoic acid), emulsifiers and anticaking agents. For each of these ingredients, safety and efficacy along the food chain needs to be assured (Fig. 1).

As an example, aflatoxin B1 present in grains fed to dairy cows may end up as aflatoxin M1 in milk products consumed by humans. Likewise, *Salmonella* present in chicken feed could mean that eggs are unsafe for consumption if contaminated with this bacterium. As well as issues of safety and quality, a life-cycle approach to feed production must also include other aspects such as workers' safety during feed processing and potential environmental pollution. A holistic feed-safety approach starts at the very beginning of the production process of a feed ingredient; continues with the investigation of the ingredient's properties and its behaviour when mixed into feed; and assesses safety for the worker exposed to the ingredient, the safety of the target animals when ingesting it, environmental safety when digested feed materials are excreted

Primary production, extraction, synthesis

↓

Feed mill, processing, storage, transport

↓

Animal, farm, stable, pasture, cage, home

↓

Meat, milk, eggs, fat, wool, ...

↓

Consumer

Figure 1 Fate of feed ingredients

by the animal, and eventually the safety of the consumer who is possibly exposed to feed constituents, their metabolites, or feed contaminants when consuming animal products.

The extent to which these aspects are covered by specific regulatory requirements differs. However, any feed producer must take legal responsibility and is liable for the wholesomeness and safety of the feed marketed. Hence the development of a new feed product, whether a single ingredient or a ready-to-consume feed, should take into consideration all aspects of quality, safety and efficacy across the life-cycle of the product.

3 Safe feed and safe food

Animal feed is an important element of the human food chain. In one way or another most feed ends indirectly on the plate of the consumer. Several food hazards originate from animal feed. Some of these resulted in a number of prominent food scares in the European Union between 1995 and 2001. Among these were incidents involving BSE, mycotoxins, dioxins, residues of banned veterinary drugs, and fears concerning the development of antimicrobial resistance. These food scares were a major driver for new food legislation in the European Union which integrated fully all aspects of feed production and animal feeding into one horizontal food safety framework (EU, 2002). This framework applies not only to feed for food-producing animals, also to feed for companion animals or for non-food-producing animals (e.g. in zoos). Until that time the safety of animal feed was seen as requiring less rigorous safety regulation compared to requirements for food for direct human consumption.

The new EU food legislation meant that the safety of feed for animals now attracted the same level of scrutiny as the safety of food for consumers in many respects, resulting in a significant increase in the data needed for feed products requiring authorisation. This increased scrutiny of safety issues relating to animal feed was accompanied by the development of guidance documents and new scientific evaluation procedures by the FEEDAP panel, the expert group at EFSA responsible for the evaluation of feed additives. The revised food law of 2002 triggered a re-evaluation of all feed additives that had been authorised by the European Union after 1970. This included a large group of flavourings with several thousand substances listed. This re-evaluation started in 2010 and created a considerable number of case studies of specific problems which allowed the experts (and the applicants) to collect additional information on the range of issues relevant to the life-cycle of a feed ingredient. Based on this information, the EFSA revised the relevant guidance documents between 2017 and 2019 and developed a set of up-to-date guidelines for the evaluation of feed ingredients (EFSA 2017a,b,c, 2018a,b, 2019a).

4 Where do you start?

As discussed, we will, for the purpose of this chapter, describe the data needed for a feed additive dossier in the European Union. When developing a dossier for a feed product, the first step is to identify the legal category it will fall into, and to understand the requirements for that legal category in the jurisdiction of choice. In the European Union feed ingredients may be either feed additives which require premarket authorisation, or feed materials which may be put into circulation without such authorisation. A health claim may be developed for a feed material or a compound feed. No authorisation is required but a dossier documenting the claim is needed which must be shared with the authorities if they request it. In addition such data may be necessary for an effective marketing strategy. In principle, similar datasets are needed in other jurisdictions such as Australia, Canada, Japan and the United States, though the emphasis of the dossier may differ.

Once you have identified the legal category and you have decided your product is a feed additive, you must draft the proposed conditions of use which will be part of the authorisation process. In that sense you start at the very end of the project (i.e. the intended use). The nature of the eventual authorisation is what drives the need for data. The authorisation of a feed additive will comprise characterisation of the additive, its function in the feed or in the animal, the categories of animals that shall receive feed containing the additive, the duration of administration, levels of the additive to be incorporated into feed, analytical methods for the additive present in the feed, and other relevant aspects, for example interaction of the additive with other feed ingredients and feed components.

All key data that support the feed product to be authorised need to be generated with the final product in mind. If studies and data were originally generated for a different product (different degrees of purity or active principle), specific arguments need to be presented that justify their use for the product that is subject of the application. These aspects of product development are familiar to those who develop pharmaceutical products, where standard project schemes are applied for single phases of product development. However, projects for new feed ingredients are often managed ad hoc without taking full account of the eventual regulatory framework and its requirements.

Another important aspect of developing a new feed product or a new use for an existing feed product is understanding the baseline of current knowledge. For this purpose, a systematic literature review is recommended that makes it possible to identify and collect all data that would be potentially relevant when discussing the application with experts at the relevant authority. For literature-based dossiers, for example for health claims, the literature review may even be the main part of the dossier, and corresponding guidance for undertaking such work is available (Glanville et al., 2014). In the context of food safety and of the regulatory framework for the evaluation and authorisation of a feed product, a literature review must be free from any bias and provide an objective summary of the already available data. Such literature reviews may be valuable steps in the generation of data on the characterisation, the safety, and the efficacy of a feed additive, and the claims for a feed product.

5 Building the dossier

Regulatory authorities provide templates for the submission of data, now usually in electronic form. These dossiers support the application of a feed ingredient for market authorisation. In addition to the template, guidelines are issued for the main chapters of a dossier. Both the template structure and guidelines allow feed producers to develop a project plan that comprises all data that need to be generated, the experiments that need to be run, and further additional work required. The data needed can be put into a logical sequence of experiments and studies. As an example, for any studies with animals at least one analytical method is needed to determine the quantity of the additive in the feed. The development of a suitable analytical method therefore needs to precede testing the additive with animals. Likewise, the safety for target animals depends on the intended use level. You first need to identify (using a dose range finder) the efficacious level of the feed product before you can test multiples of this level of feed for the safety of the target animal. Another example is consumer safety which depends on potential residues in animal products. Residues could result from the ingredient itself or after metabolization by the animal. As a result, investigations into the metabolism of the feed additive by target animals must precede the discussion of consumer safety. In reality, the picture is more complex, as data are often generated in order to justify the development of a product, which means that frequently efficacy data are produced at an early stage. However, one may then need to differentiate clearly between data generated to prove the concept of a product idea and data generated for a regulatory purpose.

For efficient project management, a right-first-time approach is advisable, as applied for human pharmaceutical products. All steps of product development are completed to a standard that can be included directly in the dossier. This minimises repeating experiments because they have not met regulatory standards and accelerates the assembly and completion of the final dossier. It is important to note that regulatory authorities discourage submission of incomplete datasets, a practice which may be characterised by the attitude of 'wait and see whether the experts say this data is are really needed'. The EFSA, for example has a unit dedicated to checking the completeness of a dossier as described in the relevant guidance documents. Incomplete dossiers are considered invalid and are therefore rejected (EFSA, 2007).

6 Characterisation and properties

The chemical composition of a feed product informs us directly about the potential hazards that may occur during the product's life cycle. Toxic impurities could endanger the health of the animal, proteins present could trigger allergenicity in workers, active substances may be sensitive to oxygen and transform into products that are less active or toxic, solubility in water may be relevant to whether the excreted additive could end up in groundwater.

First, the feed additive's composition needs to be fully characterised (EFSA, 2017a). This may be easy for substances of high purity such as crystalline vitamin C but may be a challenge if a more complex extract from a plant is the material of choice. In such cases a systematic literature review of possible constituents makes it possible to narrow down the substances that need to be characterized. In addition to the active substance that is responsible for the effects of the feed product, all other major and minor constituents typically or potentially present need to be analysed. This obviously depends on the raw material, the manufacturing process and eventually the purification steps. Certain impurities, in the European Union called 'undesired substances', need to be tested by default. Examples include dioxins in products of mineral or industrial origin, mycotoxins in products derived from plant materials, organic solvents in extracts and heavy metals in all feed products.

Besides composition other properties need to be investigated. Degradation products need to be part of stability studies since they may constitute impurities to be accounted for as part of the specification of the product. Other physicochemical properties that need to be studied are particle size, dusting potential, solubility in water and organic solvents, stability under normal and exceptional storage conditions and reactivity with other components. For specific product categories such as microorganisms, enzymes and plant extracts, separate additional datasets are needed.

For microorganisms, whether used as feed ingredients or in the production of feed ingredients (e.g. in fermentation processes), there is a need for data that address issues such as taxonomy (i.e. unambiguous identification of the species), the risk of antimicrobial resistance or the production of toxins. This is now often most easily achieved by investigating the whole genome sequence (WGS) of the organism which makes it possible to allocate the organism to its proper taxonomic species and to search for genes that indicate the presence of antimicrobial resistance or metabolic pathways leading to the formation of toxic compounds.

Secondly, data are needed to understand the properties of the product when added to feed. The most obvious properties are mixing behaviour and stability. Experimental evidence is needed that the product can be mixed homogeneously into feed of different types for example mashed or pelleted feed. The stability of the ingredient needs to be investigated during all processing stages including mixing, pelleting (at elevated temperatures) and storage of the feed. Should an additive be used in an intermediate product, for example as a premixture (which is a commercial product in its own right), data may also be needed for the premixture. All such data should reflect commercial practice. As an example, stability in premixtures should be studied for a period of 6 months, whereas stability in feed for food-producing animals can be studied for 3 months only. The required period of study depends on the expected storage recommendation for the final feed product. In the case of pet food, guaranteed stability is frequently for 1 or 2 years, and any ingredient that serves a nutritional need or is designed to have specific health effects needs to provide those benefits until the end of the 'best before' period (which is authorised for the inclusion on labelling).

Product characterisation requires analytical methods to determine each ingredient and confirm that it is present in animal feed. Analytical methods may also be needed to determine residues in animal tissues (e.g. muscle, fat or liver) and animal products (e.g. eggs or milk). In case an analytical method mentioned in the authorisation is not an official method, it needs to meet criteria that indicate that it has been sufficiently validated (EU, 2008a: Section 2.6. in Annex II).

7 Safety data: animals, consumers, workers and the environment

The first step in any safety assessment for a feed product is an inventory of available data to allow the development of a preliminary toxicological profile. This will help identify the need for new data and the sequence of experiments and studies to be undertaken to provide that data. As feed products for food-producing animals enter the food chain, some basic questions need to be addressed and understood before undertaking studies that involve major costs and resources. Substances added intentionally to the food chain must not be genotoxic, must not carry antimicrobial resistance genes or enable their occurrence and should not have hormone-like activities which might classify them as endocrine disruptors. When developing new feed products these risks should be investigated at the beginning of the process. For many processed feed ingredients these questions may be easy to answer, but natural extracts for example are complex and may contain potentially genotoxic compounds, microorganisms which may carry and spread antimicrobial resistance genes, and molecules able to bind to hormone receptors.

Once these basic questions have been addressed, the further need for classical toxicological studies depends on the composition of the compound and the chemistry of its constituents. For substances to which consumers may be exposed, the EFSA usually requests a minimum 90-day oral feeding study in laboratory rats (EFSA, 2017b). Over the period of 90 days the test substance is added to the diet of animals, usually being allocated to a control group and three groups with increasing doses (e.g. 0, 100, 500, 1000 mg/kg bodyweight per day). This study serves two purposes: to identify organs, tissues and physiological parameters that may be affected or damaged by the substance, and to assess the dietary levels at which such effects are observed or not yet observed using a dose-response approach. The level at which no adverse effects are observed is used as the point of departure to establish safe intake levels for humans and, where appropriate, animals. As an example, at the doses quoted above we might see increased liver weight in rats at 1000 mg/kg bodyweight per day, but we do not make this observation at the medium dose level of 500 mg/kg bodyweight per day. The value of 500 mg/bw*d is determined to be the 'no observed adverse effect level' (NOAEL); it is divided by a default factor of 100 to establish the intake by humans considered to be without appreciable risk. The acceptable daily intake (ADI) would in this case be 5 mg/bw, which translates for an adult of 60 kg bodyweight to a daily intake of 300 mg.

This approach of using toxicological studies to establish an acceptable daily intake for consumers was developed more than 60 years ago. The European Food Safety Authority (EFSA) has recently extended this concept to animals, whether food-producing or pet animals, applying the same default factor of 100 when deriving acceptable intakes for a broiler or dog. It is possible to establish acceptable feed levels for all animals in a species using one study in rats. However, this approach requires rather high no observed adverse effect levels in a laboratory animal study. Table 2 compares the corresponding maximum safe feed levels for two feed additives, a natural and a synthetic flavouring, as calculated by the EFSA.

It is clear from both examples in this table that substances with a low safe level in rats will lead to low save feed levels for the target animals in question. Should the intended and technologically justified level in feed be higher, the safety of the feed product needs to be studied in the target animals themselves. The strategy for safety testing in target animals will depend on the range of animal species and categories (e.g. age range) for which the substance shall be added to feed. If just one animal category is involved, for example piglets, safety is studied in this category.

Table 2 Maximum safe feed concentrations for selected target animals

Feed additive	Essential oil from Origanum vulgare ssp. hirtum (Link) letsw.	1-Octen-3-ol
Reference	EFSA, 2019b	EFSA, 2020
NOAEL (mg/kg bw*d)	200	6.7
Maximum safe concentrations (mg/kg feed)		
Chickens for fattening	22	1.1
Piglets	40	1.3
Cattle for fattening	88	2.9
Dogs	106	3.5

If the additive is to be used in a broader range of animal species a specific strategy needs to be developed. The EFSA recently systematised safety assessments by taking into account the life stage (e.g. young/growing, adult/reproductive) and the broad classifications of animals as mammals (ruminants, monogastric mammals), birds (poultry) and fish (EFSA, 2017c). A safety study in piglets is applicable to young pigs in the fattening stage, but not to adult sows used for piglet production. If an ingredient is to be used in all pig categories, separate safety studies in piglets and sows are needed. Similarly, in the case of ruminants, two separate studies in calves and in dairy cows are required. In the case of poultry, separate studies in broilers (for meat) and laying hens (for eggs) are needed. If the intention is to develop a feed additive for all animal species and categories (including pets), a minimum of four studies in piglets, broilers, dairy cows and one fish species is needed. These general principles of extrapolation between growing and reproductive animals are applied on a case-by-case basis for specific products in particular situations and should not be taken as a default as sufficient in all cases.

Safety studies in target animals are called tolerance studies. They are not to be compared with toxicological studies in rats or other laboratory animals for which detailed and extensive guidelines by the OECD are available. The EFSA has developed a set of principles to be considered for tolerance studies (EFSA, 2017c). The design of the study is driven mainly by the margin of safety that is envisaged: beside a negative control group, the ingredient is given at the highest proposed dose to the treatment group, and multiples of this dose are administered to one or more tolerance groups. Should the tolerance be hundredfold higher, a simple study observing animal growth and performance plus health monitoring is sufficient. Where the tolerance group needs to be administered a dose at a factor of 10 or more than the highest proposed dose, detailed data including monitoring of blood chemistry and haematology are needed. At tolerance levels below 10 it is expected that selected representative animals from all dose groups are subjected to histopathological examination.

Specific attention needs to be paid to ensure a sufficiently robust statistical methodology for the study which must include statistical evaluation and comparison of the parameters observed in the various groups of animals. The EFSA formally requests that the protocol for a tolerance study is a signed document that is added as an annex to the final study report. It is a good idea, as is practiced in GLP studies, to document any change with respect to the protocol and to formally sign off those changes. It is important to note that following Good Laboratory Practice (GLP) is essential if the study is intended for a product dossier for the FDA. GLP is recommended by the EFSA but not mandatory.

In assessing the safety of the feed additive for consumers, the first question to be answered is whether the feed ingredient or its metabolites will be carried over into food products. In case of colourants this carryover may be intentional (e.g. to give egg yolk or salmon a desirable appearance). For the assessment of these residues basic toxicological data packages are necessary, with a minimum requirement for mutagenicity studies and a 90-day rat study (see this section, second paragraph). The acceptable daily intake (ADI) for residues must ensure their safety for the consumer.

To understand if metabolites are deposited in animal organs or tissues, specific metabolism studies may be needed. No specific designs and protocols for such studies for feed ingredients are available since different types of materials may need to be studied. The guidance provided by the VICH is an appropriate starting point, suitably adapted to the properties of the substance being studied (EMA, 2011). Clearly plant ingredients consisting of typical nutrients or fibres eaten by animals will not raise concerns. There may also be data from laboratory animals or humans that can be used. However, since no general approach is available, investigations need to be developed on a case-by-case basis. They will depend on the data available for similar substances in predicting likely metabolism. One obstacle is that metabolism of chemicals is often only studied in laboratory animals (rats, mice or dogs), raising the question of whether the same metabolic pathways are present in other animal categories such as birds or fish.

Data on worker safety are needed in the evaluation of feed additives. Workers at feed ingredient production sites, feed mill workers and farmers may be exposed to the feed product, on the skin, in the eyes, the oral cavity and the gastrointestinal (GI) tract or by inhalation into the lungs. Corresponding guidance is available from the EFSA (2012). Of specific interest is accidental exposure at higher levels and a set of OECD study protocols are available to investigate this. Important considerations include whether the substance is solid or liquid, whether it generates dust that may be inhaled and whether it contains specific substances of concern such as endotoxins. Within the European Union any ingredient that is marketed between businesses needs to be assessed and classified for its safety with respect to workers and the environment, and needs to be labelled accordingly using the appropriate warning statements and pictograms (EU, 2008b).

Feed ingredients pass through the GI tract of animals and are excreted as such or in a modified form. They may also be spread as manure on farmland. In aquaculture feed ingredients may leak out of feed pellets or be excreted by fish or shrimps into water and end up in the sediment of a lake or below a sea cage used to contain the fish being farmed. Environmental safety evaluations of feed additives follow the approach for pesticides and veterinary drugs. The requirement may be waived if the substance occurs naturally in the environment (this excludes ingredients derived from exotic species) or is fully metabolised and not excreted at all. Hazards to be investigated are toxic effects on organisms in environmental compartments where the feed additive could be present: soil, groundwater, surface water (lakes, rivers), freshwater sediments, seawater or sea water sediments. A full evaluation requires a considerable amount of data. However, it is also possible to model the behaviour of substances in different conditions. Extensive guidance is available from the EFSA to help develop a product-specific strategy (EFSA, 2019a).

8 Effects and efficacy

Feed ingredients are developed for a wide variety of reasons. They may affect the properties of feed (e.g. a preservative will allow storage for a given period), change the behaviour of animals (e.g. an appetiser may increase feed consumption), improve gut health (e.g. pre- and probiotics), support digestion of feed constituents (e.g. enzymes), improve productivity parameters such as feed conversion ratio (e.g. increase of milk output per dry matter input), reduce environmentally hazardous materials (e.g. phosphate in faeces or exhaled methane), change the appearance of animal tissues and products (e.g. colouring egg yolks) and provide macro and micro nutrients (e.g. carbohydrates in corn flour or selenium-enriched yeast). There is no single way of studying and proving the efficacy of such a broad range of functions and effects of feed ingredients.

The number of studies required, their design and the endpoints to be investigated all depend on the specific target effect of the specific substance (EFSA, 2018a). The common denominator is that evidence of efficacy needs to be generated using scientific protocols that reflect state-of-the-art knowledge and methodology. The same general considerations apply as for animal safety. Efficacy may differ for each target animal category i.e. ruminating and monogastric mammals, poultry, fish and non-vertebrates. It may also differ for growing and reproductive animals. An ingredient that can be used for different species will require separate studies for each species. Where an ingredient is targeting the feed rather than the animal (e.g. a preservative), it must be tested on the different types of feed to which it will be added.

In the European Union, a specific group of feed additives are classified as so-called zootechnical additives, that is additives designed to optimise the performance of healthy animals. Their authorisation resembles that for veterinary products. Authorisation is holder specific and depends on "guaranteed" effects on performance, growth and health of the animal. Three independent efficacy studies are needed for each target animal category in order to demonstrate that beneficial effects are likely to be achieved regularly in the marketplace. It is important to justify the intended level of use, and an appropriate dose range finding study should be included in the study programme. Such a study may serve as one of the three required efficacy studies. Study planning also needs to take into account the number of animals required to provide statistically reliable results for both tolerance and efficacy studies.

As European agricultural practices are diverse, studies should be carried out in at least two different locations. Studies from other countries (e.g. the United States) can be accepted, but only if feed compositions and animal husbandry practices are sufficiently similar to those in the European Union. Diets offered to broilers or pigs may differ significantly between the US and Europe with respect to major ingredients used as sources for carbohydrates and proteins. Efficacy data generated in the United States may not be applicable to the European Union if such differences occur (and vice versa).

There is a considerable difference between the European Union and United States with respect to the planning of efficacy studies. The EFSA in the European Union provides detailed written guidelines and expects the applicant to read them and then prepare a final report for assessment by the EFSA (with limited opportunity to prior consultation). In contrast the FDA in the United States encourages companies to approach FDA experts and to discuss with them all relevant aspects of study protocols up front. There is a formal review process for draft study protocols which results in detailed written comments supported by further clarifications in meetings. This applies specifically to the application procedure for food additives (for animal feed) but can also be used for the generally recognised as safe (GRAS) routes.

Although it is difficult to provide extensive guidance on efficacy studies, as the endpoints of efficacy differ significantly, some elements are common to all studies. Animal studies must record all details with respect to origin of animals (health status), feed composition, analysis of ingredients in feed, dosing of the feed ingredient, feeding regime, husbandry conditions in general, specific husbandry practices (e.g. layout of stables and pens), water supply any diseases and medication used to treat them. Animals should be in good health throughout the study and receive an appropriate diet in accordance with recommendations by breeders.

It is widely accepted that it is difficult to establish efficacy under optimal conditions which may not reflect real commercial conditions. However, since findings may not be reproducible, field studies under practical conditions are not accepted by EFSA. Results from such studies may be used to corroborate the dossier, but are not considered to provide the pivotal data that allow a decision to be made about the effects and the efficacy of a feed additive.

9 Substantiating claims

In the previous section we discussed those feed ingredient effects that are assessed by an authority as part of pre-market authorisation. Some claims about the effects of a feed ingredient or feed product (including a complete feed/ complementary feed) can be made outside the regulatory procedure required for authorisation of a feed additive. Such claims are accepted in the European Union for feed materials (which are not feed additives) and complete feed, and they are possible in many other jurisdictions as long as they do not mislead the buyer/user of the feed. Being not misled about a claimed effect means that the statement is true as far as common understanding among scientists applies. Such claims are often acceptable under freedom of speech rules and can be made on the product label itself as well as on leaflets, webpages, at exhibitions and in direct discussions with customers.

In the European Union the following conditions need to be met when making a claim. It should be objective, verifiable by the competent authorities and understandable by the user of the feed. In order to meet the second condition, the feed producer should be able to produce scientific substantiation of the claim that can be shared on request with the relevant authorities (EU, 2009). The regulation suggests either publicly-available scientific evidence or documented company research as scientific substantiation for a claim. No further specific requirements are given. In that sense anybody making claims about feeds is free to document them as they think best.

One important limitation is claims for veterinary medicinal products, products treating or preventing disease in animals, or products restoring, correcting or modifying physiological functions by pharmacological, immunological or metabolic means (EU, 2019). However, there is an overlap in the area of diseases caused by improper nutrition; there are diseases or disease-like conditions such as milk fever that may be addressed by dietary intervention. In the European Union, such dietary claims related to specific diseases are regulated under the feed law for specific ingredients and target animals. The recently updated list of such claims for particular nutritional uses should be taken into account when developing new claims (EU, 2020).

Two federations representing animal feed manufacturers have recently developed more detailed guidance on the evidence files that can be used to document and justify claims for animal feeds (COPA-COGECA and FEFAC, 2018; FEDIAF, 2018). In both documents the guidance for making claims is part of more comprehensive good practice covering labelling guidelines developed by the feed sectors for food producing animals and for pet animals in close consultation with government authorities. They are recognised by these authorities and are therefore a valuable and recommended point of reference.

For a specific claim to be made for a complementary feed it will be important to consider the available evidence and to phrase carefully the actual statement. There are differences between statements such as 'plant extract XYZ improves feed conversion by 15%' and 'extracts from plants used for XYZ have been studied for their possible contribution to improved feed conversion'. To substantiate the first statement requires at least one study that demonstrates significant quantifiable effects on feed conversion. The second statement refers to studies from the published literature that may be relevant for a product but does not make any claim about a guaranteed effect.

As discussed elsewhere, a comprehensive scientific literature review is always a preferred starting point. Studies relating to the effects of plant extracts in humans (e.g. for use as traditional herbal medicines) may help to support a product, providing the feed company avoids any specific claims reserved for veterinary medicinal products which require a stronger evidence base. It is possible to use results from field studies undertaken with customers or research institutions without implementing all the requirements needed for studies which need to be evaluated by authorities like the EFSA or the FDA. Field studies are typically available in larger numbers. A meta-analysis based on a number of field studies can result in a more robust data set to support a claim. It is also possible to investigate the mode of action *in vitro* and to predict effects *in vivo* without having data from studies in target animals. However, robust claims require developing study designs and protocols similar to those for feed additives described in Section 8.

10 Dossier submission and follow-up

A successful dossier needs to comply with all legislative requirements and guidelines from the relevant authorities. In addition it should be written in a concise language that is easy to understand and which is free of company and lab jargons. The narrative parts of the submission should be clear and to the point. This requirement reflects the fact that national expert bodies are composed of individuals with different technical, scientific and linguistic backgrounds. It is important that you use the terminology and definitions laid down in the legal text and guidelines of the procedure under which you will submit your application. It is always a good idea to study reports for similar products so that you can phrase your documents in a similar way to experts.

Dossiers for EFSA are now submitted in electronic form, with the current submission by CD or USB stick, to be replaced by online submission. It is important that the dossier, its main parts and the annexes and references are submitted in an 'easy-to-use' format, for example with a clearly labelled structure and the ability to move between sections and documents via hyperlinks (EFSA, 2007). Incomplete dossiers are prevented from progressing until the missing parts have been submitted. Additional questions may be raised by experts, in which case a deadline to answer them is set by the authority in question. Follow-up of such additional requests is critical and should be prioritised within the company.

The European Union adopted recently specific rules for the notification of studies commissioned with a view to support an application for authorisation of a feed additive. Such notifications will be mandatory for industry and for contract laboratories, whether they are located in the European Union or in a third country. Failure to notify will result in a default six-month delay of the evaluation of the dossier by EFSA. These new rules will come into force on 27 March 2021. Before then the EFSA will publish more detailed guidance on the notification procedure and the publicly-accessible database where notified studies need to be filed.

11 Continuously updating the file

The processes of generating evidence for the efficacy of a feed product and assembling an evidence file are not activities that should stop once authorisation is granted and the product is launched. All feed businesses have the obligation to monitor any unexpected or unwanted effects resulting from the use of their products. As science progresses, relevant research relating to efficacy should be monitored and the evidence file updated as appropriate.

In the European Union there is currently an obligation to renew authorisation of feed additives after 10 years. The dossier for a renewal application needs to be submitted one year before the end of authorisation. In practice preparation for such a dossier starts 6–8 years after the first authorisation. The focus of the renewal dossier depends on the substance category. In all cases possible changes in the manufacturing and quality of the additive need to be discussed, and the impact of any change on safety and efficacy of the additive needs to be assessed. Relevant progress in science related to the additive needs to be documented by a systematic literature search.

12 Where to look for further information

European Union	European Food Safety Authority: www.efsa.europa.eu.
	European Union Reference Laboratory for additives for use in animal nutrition - EURL for Feed Additives: ec.europa.eu/jrc/en/eurl/feed-additives.
	European Commission - DG SANTÉ: ec.europa.eu/food/safety/animal-feed_en.
United States	US Food and Drug Administration: www.fda.gov/animal-veterinary/products/animal-food-feeds.
	Association of American Feed Control Officials: www.aafco.org/.
Sector guidance	FEFAC - European Feed Manufacturers' Association: www.fefac.eu. FEFANA - EU Association of Specialty Feed Ingredients and their Mixtures: www.fefana.org.
	Fediaf-the European petfood industry: www.fediaf.org.

13 References

Codex Alimentarius Commission (2013). *Guidelines on the Application of Risk Assessment for Feed*. CAC/GL 80-2013.

COPA-COGECA and FEFAC (2018). *EU Code of Good Labelling Practices for Compound Feed for Food Producing Animals*. European Feed Manufacturers Federation.

EFSA (2007). Administrative guidance to applicants on the preparation and presentation of applications for authorisation of additives for use in animal nutrition. *EFSA Journal* 5(5): 1447. [63 pp.] https://doi.org/10.2903/j.efsa.2007.1447, Updated on 22 March 2019.

EFSA (2012). Guidance on studies concerning the safety of use of the additive for users/workers. *EFSA Journal* 10(1): 2539. [5 pp.] https://doi.org/10.2903/j.efsa.2012.2539.

EFSA (2017a). Guidance on the identity, characterisation and conditions of use of feed additives. *EFSA Journal* 15(10): 5023. [K been the 12 pp.] https://doi.org/10.2903/j.efsa.2017.5023.

EFSA (2017b). Guidance on the assessment of the safety of feed additives for the consumer. *EFSA Journal* 15(10): 5022. [17 pp.] https://doi.org/10.2903/j.efsa.2017.5022.

EFSA (2017c). Guidance on the assessment of the safety of feed additives for the target species. *EFSA Journal* 15(10): 5021. [19 pp.] https://doi.org/10.2903/j.efsa.2017.5021.

EFSA (2018a). Scientific opinion on guidance on the assessment of the efficacy of feed additives. *EFSA Journal* 16(5): 5274. [25 pp.] https://doi.org/10.2903/j.efsa.2018.5274.

EFSA (2018b). Guidance on the characterisation of microorganisms used as feed additives or as production organisms. *EFSA Journal* 16(3): 5206. [24 pp.] https://doi.org/10.2903/j.efsa.2018.5206.

EFSA (2019a). Guidance on the assessment of the safety of feed additives for the environment. *EFSA Journal* 17(4): 5648. [78 pp.] https://doi.org/10.2903/j.efsa.2019.5648.

EFSA (2019b). Safety and efficacy of an essential oil from Origanum vulgare ssp. *hirtum* (Link) letsw. for all animal species. *EFSA Journal* 17(12): 5909. [14 pp.] https://doi.org/10.2903/j.efsa.2019.5909.

EFSA (2020). Scientific opinion on the safety and efficacy of oct-1-en-3-ol, pent-1-en-3-ol, oct-1-en-3-one, oct-1-en-3-yl acetate, isopulegol and 5-methylhept-2-en-4-one, belonging to chemical group 5 and of isopulegone and a-damascone belonging to chemical group 8 when used as flavourings for all animal species. *EFSA Journal* 18(2): 6002. [16 pp.] https://doi.org/10.2903/j.efsa.2020.6002.

European Medicines Agency (2011). VICH GL46: Studies to Evaluate the Metabolism and Residue Kinetics of Veterinary Drugs in Food-Producing Animals: Metabolism Study to Determine the Quantity and Identify the Nature of Residues. 14 March 2011. EMA/CVMP/VICH/463072/2009. Committee for Medicinal Products for Veterinary Use (CVMP).

European Union (2002). Regulation (EC) No 178/2002 of the European Parliament and of the Council of 28 January2002 laying down the general principles and requirements of food law, establishing the European Food Safety Authority and laying down procedures in matters of food safety. Official Journal of the European Communities, 1.2.2002, L31/p. 1.

European Union (2008a). Commission Regulation (EC) No 429/2008 of 25 April 2008 on detailed rules for the implementation of Regulation (EC) No 1831/2003 of the European Parliament and of the Council as regards the preparation and the presentation of applications and the assessment and the authorisation of feed additives. Official Journal of the European Union, 22.5.2008, L133/p. 1.

European Union (2008b). Regulation (EC) No 1272/2008 of the European Parliament and of the Council of 16 December 2008 on classification, labelling and packaging of substances and mixtures, amending and repealing Directives 67/548/EEC and 1999/45/EC, and amending Regulation (EC) No 1907/2006. Official Journal of the European Union, 31.12.2008, L353/p. 1.

European Union (2009). Regulation (EC) No 767/2009 of the European Parliament and of the Council on the placing on the market and use of feed, amending European Parliament and Council Regulation (EC) No 1831/2003 and repealing Council Directive 79/373/EEC, Commission Directive 80/511/EEC, Council Directives 82/471/EEC, 83/228/EEC, 93/74/EEC, 93/113/EC and 96/25/EC and Commission Decision 2004/217/EC. Official Journal of the European Union, 1.9.2009, L229/p. 1.

European Union (2019). Regulation (EU) 2019/6 of the European Parliament and of the Council of 11 December 2018 on veterinary medicinal products and repealing Directive 2001/82/EC. Official Journal of the European Union, 7.1.2019, L4/p. 43.

European Union (2020). Commission Regulation (EU) 2020/354 of 4 March 2020 establishing a list of intended uses of feed intended for particular nutritional purposes and repealing Directive 2008/38/EC. Official Journal of the European Union, 5.3.2020, L67/p. 1.

FEDIAF (2018). *Code of Good Labelling Practice for Pet Food*. European Pet food Industry.

FEFAC (2019). *Feed & Food Statistical Yearbook* (vol. 2018). European Feed Manufacturers Federation.

Glanville, J., Wood, H., Arber, M., Varley, D., Frampton, G. and Brazier, H. (2014). Technical Manual for Performing Electronic Literature Searches in Food and Feed Safety. Annex to Inventory of Sources of Scientific Evidence Relevant to EFSA's Risk Assessments and Information Sessions on Literature Searching Techniques (CFT/EFSA/SAS/2011/03). EFSA Supporting Publication.

Smedley, K. O. (2013). Comparison of Regulatory Management of Authorized Ingredients, Approval Processes, and Risk-Assessment Procedures for Feed Ingredients Jurisdictions Covered: Brazil, Canada, China, European Union, Japan, South Africa, and United States. July, 2013. Published by the International Feed Industry Federation.

Index